W9-BDT-079

Why Don't Your
Eyelashes Grow?

Beth Ann Ditkoff, M.D.

with Andrea Ditkoff and Julia Ditkoff

Avery • *a member of Penguin Group (USA) Inc.* • *New York*

Why Don't Your Eyelashes Grow?

Curious Questions

Kids Ask About

the Human Body

AVERY

Published by the Penguin Group
Penguin Group (USA) Inc., 375 Hudson Street, New York, New York 10014, USA ·
Penguin Group (Canada), 90 Eglinton Avenue East, Suite 700, Toronto, Ontario M4P 2Y3, Canada
(a division of Pearson Canada Inc.) · Penguin Books Ltd, 80 Strand, London WC2R 0RL,
England · Penguin Ireland, 25 St Stephen's Green, Dublin 2, Ireland (a division of Penguin Books Ltd)
· Penguin Group (Australia), 250 Camberwell Road, Camberwell, Victoria 3124, Australia
(a division of Pearson Australia Group Pty Ltd) · Penguin Books India Pvt Ltd, 11 Community
Centre, Panchsheel Park, New Delhi–110 017, India · Penguin Group (NZ), 67 Apollo Drive,
Rosedale, North Shore 0632, New Zealand (a division of Pearson New Zealand Ltd) · Penguin Books
(South Africa) (Pty) Ltd, 24 Sturdee Avenue, Rosebank, Johannesburg 2196, South Africa

Penguin Books Ltd, Registered Offices: 80 Strand, London WC2R 0RL, England

Most Avery books are available at special quantity discounts for bulk purchase for sales
promotions, premiums, fund-raising, and educational needs. Special books or book excerpts also
can be created to fit specific needs. For details, write Penguin Group (USA) Inc. Special Markets,
375 Hudson Street, New York, NY 10014.

Library of Congress Cataloging-in-Publication Data

Ditkoff, Beth Ann.
Why don't your eyelashes grow? : curious questions kids ask about the human body /
Beth Ann Ditkoff ; with Andrea Ditkoff and Julia Ditkoff.
p.   cm.
ISBN 978-1-58333-323-5
1. Human physiology—Miscellanea.   2. Human body—Miscellanea.   I. Ditkoff, Andrea.
II. Ditkoff, Julia.   III. Title.
QP38.D48   2008   2008027061
612—dc22

Printed in the United States of America
7   9   10   8   6

BOOK DESIGN BY NICOLE LAROCHE

For Charlie

CONTENTS

Introduction   xv

Your Body 101

Why do you have bumps on your tongue?   3

Why do you have earwax?   4

What is that dewdrop thing in the back of your mouth?   5

Why can't human beings breathe underwater?   6

Why is your eye color blue, brown, or green?   7

Why do you have eyebrows?   8

Why do you hiccup?   9

Why are the ends of your nails white and
    the other part pink?   10

Why do you have wisdom teeth?   11

Why can't your skin be purple, green, or blue?   12

Why do you have an appendix?   13

Why do you have baby teeth when you're a child?   14

Why are your lips red?   15

What happens when your legs or arms "fall asleep"?   16

What is the difference between identical and
    fraternal twins?   17

Why does your stomach growl? 18

Why do people snore? 19

Why are you left-handed or right-handed? 20

Why do we get cavities? 21

Why are we ticklish? 22

Why do some people have straight hair and
some people have curly hair? 23

Why do we blush? 24

Why do we yawn? 25

What is your Adam's apple? 26

Why do we get scabs? 27

Why do some people have crooked teeth and
others have naturally straight teeth? 28

Why do you get goose bumps when you are cold? 29

What is the black circle in your eye? 30

Why is blood red? 31

What is your ear made of? 32

Why does your blood look blue when it is in your body? 33

Why do I laugh when I think something is funny? 34

Why do human beings cry? 35

What are moles? 36

Why do we form scars? 37

What do your tonsils do? 38

How does food give you energy? 39

Why do you have a belly button? 41

Why does your mouth make spit? 42

What's a concussion? 43

Why does your heart make a "lub-dub" sound
    when it beats?   44

What is a charley horse?   45

What is a cell?   46

Why is a bruise black and blue?   47

Why do some people have freckles?   48

Why do some people have dimples?   49

What are genes?   50

What happens when you break a bone?   51

The Weird, the Ugly, and the Downright Gross

Why don't your eyelashes grow?   55

Why doesn't it hurt when you get a haircut?   56

Why do some people have "innie" belly buttons
    and some people have "outies"?   57

Why do you see fireworks when you close your eyes
    and press on them?   58

Why do we get wrinkles as we get older?   59

Why do we burp?   60

Why is your poop brown?   61

Why do we get gray hair?   62

Why do we sometimes feel nauseated while
    in a car or on a bus?   63

What are lice?   64

Why do we have nightmares?   65

Why do toenails grow slower than fingernails?   66

Why do people get pimples? 67

Why do we have boogers? 68

Why do people have hair in their nose? 69

Why do farts smell? 70

Why do you get bad breath? 71

What are warts? 72

Can someone really have six toes on one foot? 73

Why don't human beings have tails? 74

Why do your teeth turn yellow? 75

Why do some people have bad body odor? 76

Why do your fingers and toes look like prunes
  when you come out of the pool? 78

Why are some people able to wiggle their ears? 79

Why is pee yellow? 80

What are bedbugs? 81

Why do some women have mustaches? 82

Why do some men have boobs? 83

Why do some people have fat that shakes like Jell-O? 84

Why do people vomit? 85

What does it mean to be double-jointed? 86

Why do people bite their nails? 88

Why do tears taste salty? 89

What are growing pains? 90

Why do babies have a "soft spot" on their head? 91

Why can't one eye move up while the other
  eye moves down? 92

Why do your eyes close when you sneeze? 93

Why do men pee standing up and women
    pee sitting down?   94

Why can't men become pregnant?   95

Why do some men go bald?   96

What does it mean to be tongue-tied?   97

Why do people get dandruff?   98

Body Afflictions and Everyday Strange Things That Can Happen

Why are some people color-blind?   101

What makes some people stutter?   102

Why do some people need to wear eyeglasses?   103

Why do some people have food allergies?   104

Why, when I'm asleep at night, do I sometimes
    startle myself awake?   105

Why do your ears pop when you're on a plane?   106

Why do you get sunburned?   107

Why do you sometimes get a stitch in your side
    when you run or walk fast?   108

Why do people sleepwalk?   109

Why does blond hair turn green in a pool?   111

Why does your heart beat faster when you exercise?   112

What is athlete's foot?   113

Why do your lips turn blue when you swim
    in cold water?   114

Why do mosquito bites itch?   115

Why do some people grind their teeth at night?   116

Why does your face pucker when you suck
    on a lemon?   117

Can your contact lens get lost in your eye and
    float into your brain?   118

Why do some people get fat?   119

Why do mosquitoes bite some people and not others?   120

What is a hangnail?   121

Why can't I get rid of this song that is stuck
    in my head?   122

Why does poison ivy make us itch?   124

Urban Myth and What If?

Is it really bad to crack your knuckles?   127

What happens if you swallow a penny?   128

What happens if you eat dog food?   129

Can you tell your fortune by reading your palm?   130

Why do you get a headache if you eat ice cream
    too quickly?   131

Can you really get rabies from a dog bite? What is
    rabies, anyway?   132

What happens if you swallow gum?   133

If you put a pea up your nose, will it go into
    your brain?   134

Can you catch a cold if you don't wear a coat
    in winter?   135

Can your MP3 player really cause you to go deaf?   *136*

What would happen if I never brushed my hair?   *137*

If you drop food on the floor, is it still safe to eat
    if it doesn't touch the floor for more than
    five seconds?   *138*

Is it true that a dog's mouth is cleaner than
    a human's?   *139*

If you eat Pop Rocks candy and drink soda at
    the same time, will your stomach explode?   *141*

If you eat carrots, will you see better?   *142*

Is eating raw cookie dough really bad for you?   *143*

Is it true that if you swallow a watermelon seed,
    you will grow a watermelon in your stomach?   *144*

Do twins have the same fingerprints?   *145*

Can you get lead poisoning from a pencil?   *146*

Can you be allergic to jewelry?   *147*

Are green bananas bad for you?   *148*

Do you really need to drink eight 8-ounce
    glasses (8 x 8) of water a day to be healthy?   *149*

Can you get frostbite in just a few minutes if it's
    really cold outside?   *150*

Will you go blind if you sit too close to the TV?   *151*

Why is breakfast the most important meal?   *152*

Does a tongue piercing hurt?   *153*

If you cross your eyes will they get stuck?   *155*

Is caffeine bad for you?   *156*

Can eating chicken soup really cure a cold?   *157*

Does an apple a day really keep the doctor away?   158
Do you really need to wait an hour after eating
    before going swimming?   160

Bonus Body Trivia

How does an X-ray work?   163
What does an IQ test measure?   164
How much does your brain weigh?   165
How many bones are in your body?   166
What is the smallest bone in your body?   167
What is the smallest muscle in your body?   168
What is the longest muscle in your body?   169
How many times a day do you blink?   170
How many times a day does your heart beat?   171

INTRODUCTION

As a physician, I am used to answering medical questions from my patients. However, friends and family also like to ask me health questions, ranging from serious topics like brain tumors to trivial ones like ingrown toenails. I imagine if I had become an accountant, neighbors would ask me about tax problems, or if I had gone to law school, acquaintances would stop me at the grocery store to ask for legal advice.

My two daughters, Andrea and Julia, also like to query me about medical matters. One day, my husband, Charlie, heard the girls ask me, "Why don't your eyelashes grow?" It was a basic question, but one for which most people don't know the answer. As I explained to the girls about the cycle of hair growth, it gave my husband an idea. Why not write a book about the answers to everyday medical questions? Andrea and Julia would compile the questions, and I could provide the answers.

Within days, Andrea and Julia had come up with dozens of questions—involving human body trivia (How many bones are in the body?), gross scientific facts (Why do we have boogers?), urban myths (Will you go blind if you sit too close to the TV?), and basic medical facts (Why do some people have innie belly buttons and others have outies?). Soon the girls' schoolmates and friends and relatives were coming up with questions. Even our family dog, a shih tzu–poodle mix named Bob Dylan, inspired a question about what happens if you eat dog food. In a little more than a month, we had more than 150 questions to answer!

This book is intended for children and parents, to address the everyday medical questions that come up frequently in day-to-day life. We enjoyed writing it, and we hope you have fun reading the explanations for some very basic body and human health subjects.

Why Don't Your
Eyelashes Grow?

Your Body 101

the thighbone's connected to the knee bone. The knee bone's connected to the shinbone. . . ." We've all sung that song, and we all know the basic anatomy of the human body. But in this chapter, we try to get down to specifics. We look in every nook and cranny of the body to find out about earwax, the bumps on your tongue, and that dewdrop thing in the back of your throat. We search for answers about the way your body works—why are some people left-handed and some are right-handed? And we explore strange noises—like stomach growling and snoring. This part provides answers to all of your basic human body questions.

Why do you have bumps on your tongue?

the little bumps on your tongue are called taste buds. The easiest ones to see are those in front—they look like little tiny mushrooms and are called fungiform papillae. Other types of taste buds, designated vallate and foliate, are more difficult to see. Taste buds are present not only on your tongue, but also in the back of your throat. They are responsible for five different types of taste: sweet, sour, bitter, salty, and metallic (meaning it tastes like metal). Recently, researchers have added an additional taste recognized by your taste buds, known as umami. This taste is described as savory. Examples include aged cheeses, portobello mushrooms, and smoked or cured pork or fish. Your taste buds do not work by themselves—they work with your nose, which provides a sense of smell. If you hold your nose and eat a chocolate cupcake, you may not be able to tell that it is chocolate-flavored rather than vanilla. That's why when you have a cold and your nose is stuffed up, your food can taste very bland.

In the middle of your tongue and toward the back (stick your tongue way out and look in the mirror) are tiny little rows of rough lines. These are called filiform papillae, and their texture helps you lick soft foods like ice cream. If you touch the back part of your tongue, you will be able to feel these rough threads or folds.

Why do you have earwax?

the medical term for earwax is cerumen. We all have earwax, but some people have more than others. The wax is dead, dried skin cells combined with secretions from your ear skin. The wax in your ears is there for a reason—it traps dust and dirt and prevents it from falling onto the eardrum (the delicate structure that catches sound vibrations and helps you hear). There are two types of earwax—wet and dry. The wet wax is usually light or dark brown and is sticky. The dry kind is whitish-gray and is usually flaky.

Earwax builds up and then usually washes out of the ear during your bath or shower. If you push the earwax into your ear with a cotton swab, the earwax will build up and become impacted, or turn hard. This plug of wax can interfere with your hearing or irritate the lining of the ear canal and lead to infection of the outer ear and ear canal (otitis externa).

What is that dewdrop thing in the back of your mouth?

IIIIIIIIIIIIIIIIIIIIIIIIIIIIIIIIIIIIIIIIIIIIIIIIIIIIIIIIIIIIIIIIIIIIIIIIIIIIIIIIIIIIIIIIIIII

t hat thing is called the uvula. After food is chewed, but before it is swallowed, it passes through the back of your throat, called the pharynx, which is divided into the nose part of the pharynx and the throat part of the pharynx. The dewdrop uvula prevents food from coming back out through the nose by pushing it into the back of the throat and down the esophagus on its way to the stomach. If you are laughing when you drink your milk, the uvula cannot do its job properly—and milk will come out of your nose!

Why can't human beings breathe underwater?

When you breathe in air, the air travels from your nose, down your trachea (windpipe), and into your lungs. As the lungs branch into smaller and smaller airways, they end in specialized sacs called alveolae. Here, oxygen passes through the lung membranes into the bloodstream, and waste products like carbon dioxide flow out of the blood and into the air, and are subsequently expelled when you breathe out.

Fish also need oxygen to live, but their lungs are not designed to extract oxygen from the air. Instead, by passing the water through their specialized organs (called gills), they can remove the oxygen and eliminate waste gases. Since humans do not have gills, we cannot extract oxygen from water. Some marine mammals, like whales and dolphins, do live in water, but they don't breathe it. They have developed a mechanism to hold their breath for long periods of time underwater. Eventually, however, they have to come to the surface to exhale and then take a new breath.

Why is your eye color blue, brown, or green?

||||||||||||||||||||||||||||||||||||||||||||||||||||||||||||||||||||||||||||||||||||||||||||||||||||||

to answer this question, you have to do a quick review of Gregor Mendel's work in the 1800s. Mendel, who was a monk, is often called the father of genetics because of the work he did in his garden performing experiments on pea plants. Mendel discovered that there are specific genes that determine the shape of a pea seed. A gene is code that determines what a plant, animal, or person will look like. For each trait, such as eye color, we receive one gene from our mother and one from our father. So if a person receives a brown eye gene from their mother and a brown eye gene from their father, that child will have brown eyes. Similarly, if a person receives a blue eye gene from their mother and a blue eye gene from their father, that child will have blue eyes. But if a child receives one brown eye gene and one blue eye gene, that child will have brown eyes because the brown eye gene is dominant over the blue eye gene, which is recessive.

Because there are more than two genes for eye color, there are many variations of color, such as green and hazel. Rarely, someone can be born with two different-colored eyes; this condition, called heterochromia, is due to a lack of color in one eye, which makes that eye appear lighter than the other. Heterochromia can also occur after an eye injury, which can lead to a loss of pigment in the damaged eye.

Why do you have eyebrows?

We have eyebrows for the same reason we have eyelashes— to prevent dirt and specks of dust from getting into the eye. If a small particle of debris does get into the eye, we have an additional protective mechanism. The eye creates tears, and the eyelids blink automatically as a reflex to wash away the dirt and clean the eye. Also, if you get caught in the rain, your eyebrows are designed to let the water flow around your eyes rather than let it enter them.

Why do you hiccup?

‖‖‖

t he medical term for hiccups is singultus. Hiccups are caused by an irritation of the phrenic nerve. This is the nerve that supplies the diaphragm—the muscular divider between the chest and the abdomen. The diaphragm assists in breathing by rising and falling, which decreases and increases the size of the chest to allow the lungs to expand when we take a breath. If the nerve supplying the diaphragm is irritated, the movement of the diaphragm is no longer coordinated with the chest, and hiccupping results. Fortunately, most of the time, hiccups go away on their own or with simple home remedies such as holding your breath, being startled, or drinking a glass of water. Very rarely, for reasons that we don't understand, hiccups persist and prescription medication may be needed to stop them.

Why are the ends of your nails white and the other part pink?

||||||||||||||||||||||||||||||||||||||||||||||||||||||||||||||||||||||||||||||||||||||||||

Your fingernails and toenails are made of a protein called keratin. This protein is colorless. The pink part of your nail comes from the blood vessels (called capillaries) in the skin beneath it. The red blood in the capillaries shows through the translucent keratin and appears pink. Because there is no skin or capillaries underneath the tips of your nails, just the whitish color of the keratin shows through.

Why do you have wisdom teeth?

|||||||||||||||||||||||||||||||||||||||||||||||||||||||||||||||||||||||||||||||||||||||||||||||||

When you are a baby/toddler you have twenty baby teeth. These teeth usually come in between six and twenty-four months. Then these baby teeth usually fall out from ages six to twelve years. They are replaced by thirty-two adult teeth, including both the first and second molars. These teeth usually come in by about age eighteen. The third molars, or wisdom teeth, may not come in until young adulthood—up to age twenty-five, hence the name "wisdom," because as you get older you get wiser! If these wisdom teeth are twisted or impacted they may never come in, and you might have to have surgery to remove them if there is a risk of infection or abscess formation.

Why can't your skin be purple, green, or blue?

the color of human skin and hair is determined by a pigment called melanin. The word "melanin" is derived from the Greek word *melas,* meaning "black." Melanin is found within keratinocytes, which are specialized skin cells. The amount and location of this pigment determine how dark the skin will be. People who do not have a lot of melanin have white, fair skin. People who have a lot of melanin have dark, black skin, and people who have an intermediate amount have a range of skin colors, including yellowish and reddish. When you sit in the sun, your skin melanin becomes darker and more melanin is formed, which causes a suntan. However, exposing your skin to sun without sun protection factor lotion can cause wrinkles, spotting, and even skin cancer. You should always wear sun protection when you are outdoors, especially during the peak sun hours of ten a.m. to two p.m. Your hair has a specialized melanin pigment called pheomelanin, which can make hair color red. Because there are no purple, green, or blue pigments in your skin and hair, you cannot have purple, green, or blue skin or hair unless you use makeup or hair dye.

Why do you have an appendix?

||||||||||||||||||||||||||||||||||||||||||||||||||||||||||||||||||||||||||||||||||||||||||||||||||

t he appendix is a small pouch, several inches long, that extends from the right side of your large intestine. For a long time, we thought that the appendix had no function, but more recently it has been thought that the appendix plays a role in the body's immune system. Immune tissue has been found in the appendix, similar to the tissue found in lymph nodes. Remember that lymph nodes help to fight infection—when you have strep throat, the lymph nodes in your neck swell to help defend your body from the bacterial infection. Although the appendix is helpful to the immune system, it is not essential, and there are no problems if your appendix needs to be removed with surgery.

One of the most common reasons to remove the appendix is when it becomes inflamed—this is known as appendicitis. If the opening to the pouch of the appendix is blocked by local swelling, stool, or occasionally seeds from fruits or vegetables that you might have eaten, the appendix becomes engorged and causes belly pain, fever, and vomiting. Eventually the appendix will burst, causing a more widespread infection inside the abdomen. The only treatment is surgery to remove the appendix. The most common age to get appendicitis is the mid-teens.

Why do you have baby teeth when you're a child?

the twenty baby teeth you have usually fall out before the teen years. These blunt teeth are replaced by specially shaped adult teeth designed to help you chew your food. With the exception of the wisdom teeth, there are sixteen adult teeth in the top and sixteen in the bottom jaws—including four incisors, which are sharp for cutting, two canine teeth, which are pointed for piercing, and four premolars and six molars, which are broad and flat for grinding. You need these specialized teeth to chew tough food, like steak.

Why are your lips red?

first of all, did you know that the outline or the border of your lips (called the vermillion border) is a special feature of humans only? This transition line from your skin to the pinkish-red part of your lips is found only in humans—no one knows why. The lips appear red because of the underlying blood vessels. Arteries are blood vessels that carry blood away from the heart and veins are blood vessels that carry blood back to the heart. The arteries and veins are connected through a series of tiny interlocking loops of blood vessels called capillaries. These red-colored blood-filled capillaries are close to the thin skin on your lips, so your lips appear red.

What happens when your legs or arms "fall asleep"?

||||||||||||||||||||||||||||||||||||||||||||||||||||||||||||||||||||||||||||||||||||||||||||||||||||||||

W e've all had this sensation—you wake up from a sound sleep and your arm or leg is numb and tingling. You can't move it, and it can be very painful. When you touch your arm, you can't feel it. You sit up and rub your arm and slowly it comes back to normal. The sensation of "falling asleep" is called paresthesia in medical terminology. It happens when there is decreased blood flow to the arm or leg. The nerves supplying sensation are very, very sensitive to this decreased blood flow and they give out signals of numbness, tingling, and pain. These sensations can also happen when you cross your legs or sit cross-legged on the floor. The treatment for the paresthesia is to change your position and gradually your arm or leg will come back to normal and avoid permanent nerve damage.

What is the difference between identical and fraternal twins?

|||||||||||||||||||||||||||||||||||||||||||||||||||||||||||||||||||||||||||||||||||||||||||||||

there are two types of twins, identical and fraternal. Identical twins, also called monozygotic twins, are the result of one sperm fertilizing one egg, and then that fertilized egg develops into two identical twins. These twins are always the same sex and always look alike. Fraternal, or dizygotic, twins, on the other hand, result from two sperm fertilizing two separate eggs. These fertilized eggs then grow into two babies. The babies can be the same sex or one boy and one girl. The two twins usually do not look more alike than other siblings.

If your mother or grandmother had twins, you may be more likely to have twins too. Fertility drugs (medicines to help a woman become pregnant) can also cause twins. Because of the increased use of these medicines, the rate of multiple births in this country is around 3 percent of all births. Most of these children are twins, but some can be triplets (three), quadruplets (four), quintuplets (five), sextuplets (six), septuplets (seven), octuplets (eight), or even higher. As hospital care is improving, more and more of these higher multiple birth babies are surviving and growing up to lead healthy lives.

Why does your stomach growl?

When you eat, food travels from your mouth, down your esophagus, through your stomach, into your intestines (small and large), and then out your rectum as a bowel movement. As the food moves through your digestive tract, it is broken down by intestinal fluids, and the nutrients are extracted and absorbed into the body. The intestines are a muscular tube that contracts and forces food, air, and liquid through it. After you eat, you may hear a loud growling sound. The medical term for such a sound is borborygmus (plural borborygmi). These sounds are normal and represent the food moving inside you as it is digested. Sometimes these sounds can be long, loud, and embarrassing, but they are still normal. You may also hear borborygmi when you are hungry—in this case, the gut is moving fluid and air through it, in preparation for you to eat—so if you skip breakfast, be prepared for the consequences.

Why do people snore?

||||||||||||||||||||||||||||||||||||||||||||||||||||||||||||||||||||||||||||||||||||||||||||||||||||

he back of your throat is made up of muscles. When you are awake, these muscles are alert and do their job properly— they coordinate so that food passes through the esophagus into the stomach and air passes through your windpipe (trachea) into your lungs. When you are asleep, your throat muscles relax— sometimes they become too relaxed, and they rub against each other as you breathe, creating a vibrating sound called snoring.

Snoring in children is often caused by large tonsils in the back of the throat. The older you get the more likely you are to snore— that's why when your grandfather takes a nap, you can probably hear him! Snoring is worse for people who are overweight or who sleep on their backs. If you or someone you know is a snorer, you should see the doctor to find out the cause and how to stop it. Sometimes it's as easy as losing weight or wearing a nose strip to open up the nasal passages. Other people may need to have surgery to correct the problem.

m ost people are right-handed—meaning they prefer to do things with their right side, like writing, throwing a baseball, or kicking a soccer ball. About 10 percent of people are left-handed, and an even smaller number of people can do things equally well with both hands. These people are called ambidextrous.

Your brain is divided into two halves—the right half and the left half. The right side of the brain controls the left side of your body. If you want to pick up a pen with your left hand, it's the right brain that sends the order. The left side of your brain is in charge of the right side of your body. In addition, each half of the brain is responsible for specific kinds of thinking. The right half is responsible for your artistic side—your imagination, and shapes and patterns. The left brain is responsible for your logical side. It helps you do math, learn a new language, and analyze a problem. Each person has a dominant, or most important, side. If your left brain is dominant, you will be a rightie. If your right brain is dominant, you will be a leftie.

Being a leftie or a rightie is something you are born with. When doctors look at developing babies in their mothers' bellies using an ultrasound machine, they can see babies sucking their thumb— if it's the left thumb, that baby probably will be born a leftie, same for the right thumb and being a rightie.

Why do we get cavities?

||||||||||||||||||||||||||||||||||||||||||||||||||||||||||||||||||||||||||||||||||||||||||||||||||||||||||||||||

C avities are literally tiny holes in your teeth. Slowly, with time, the holes get bigger and bigger and the inside of the tooth rots. Cavities start when you allow plaque to build up on the surface of your teeth. Plaque is that whitish germy stuff that you brush away every morning and every night. If you don't do a good job brushing and flossing, the plaque can remain on the surface of your teeth and it can start to eat away at the outer part of the tooth called the enamel. It's really just like making a hole in a tree by stripping away the bark and allowing the inside part of the tree to rot or decay.

Most people don't even know they have a cavity until they visit the dentist. After finding a cavity, the dentist will clean out all of the rotted part of the tooth and fill in the hole with some filling material (usually the white color of your tooth so it doesn't show). If you don't see your dentist, the cavity grows larger, until eventually your tooth starts to hurt or gets infected—or even worse, until it falls out. The most common teeth affected are the back teeth, because they have so many nooks, grooves, and crannies that can trap the plaque.

So it's important to brush and floss. Use toothpaste that contains fluoride and drink tap water that has fluoride in it (unlike most bottled water)—the fluoride helps to protect your teeth and prevent cavities. Also, try not to eat and drink too many sugary things like candy and juice—the sugar can encourage plaque to form and the decay process to begin.

Why are we ticklish?

||||||||||||||||||||||||||||||||||||||||||||||||||||||||||||||||||||||||||||||||||||||||||||||||||||||||||||||||||||||||||||||||||

S ome researchers think tickling serves as a bonding between mother and child. Infants will laugh when their mothers tickle them, thus serving to build the social interaction and allowing the child to interact with his or her environment.

There are two types of tickle sensations. The first is knismesis, a light movement on the skin like an insect crawling up your arm. This type of tickling is annoying. The second type of tickling is gargalesis, and it is the fun, smiling, laughing kind of tickling that everyone is familiar with. Some people are not ticklish, others are very ticklish. We don't know what accounts for these individual differences.

In 1872, Charles Darwin was able to identify the five criteria for successful tickling: (1) The tickler has to be someone you know and are familiar with. If your brother tickles your armpit, it makes you giggle; if a stranger did that, though, you would not find it funny. (2) You can't make yourself laugh by tickling yourself. Scientists think that when you try to do so, a part of the brain (the cerebellum) turns off the giggle response. (3) The body part that is being tickled is one that is not usually touched by other people. The most ticklish areas of the body are the soles of the feet, the underarms, the waist/ribs, and the area under the neck and chin. (4) The tickling has to be a light touch. If your tickle fight becomes too rough, it's not fun anymore. (5) Finally, tickling causes laughter more easily if you're already in a light and happy mood.

Why do some people have straight hair and some people have curly hair?

almost everyone who has curly hair wants straight hair and everyone who has straight hair complains that they want curly hair! Your hair is made up of a protein called keratin. This protein is formed by individual building blocks (atoms). The more connections or bonds between these building blocks, the curlier your hair is. If you have fewer bonds, your hair is straighter. Even if you get a perm to make your hair curlier or undergo a chemical process to make your hair straighter, these procedures are only temporary. The new keratin that is produced at the scalp level has the same number of bonds as before, so your new hair will grow in like it usually is—curly or straight.

Why do we blush?

||||||||||||||||||||||||||||||||||||||||||||||||||||||||||||||||||||||||||||||||||||||||||||||||||||||

blushing is the red color that appears on your face, ears, and neck caused by increased blood flow to the blood vessels in the skin. The medical term for blushing is craniofacial erythema. You blush in response to some emotional or social trigger—like being embarrassed in public or feeling ashamed when you notice that you forgot to zip up your jeans. Blushing comes in spurts and is accompanied by a feeling of warmth or heat in the face and a burning/tingling sensation. You cannot control your blushing—once it starts, you can't stop it.

Although the redness from blushing can be seen best in fair-skinned people, it has been reported in all races. Blushing can be a symptom of emotional problems, medication side effects, or skin conditions, but most often there is no underlying cause. Blushing can be covered up with makeup. Sometimes people undergo behavioral treatment or take medicine (to decrease the heart rate and calm the nerves) to try to stop the blushing. In extreme cases, when people blush all the time, surgery has been used to destroy the nerves that cause the increased facial blood flow response in order to eliminate this embarrassing social problem.

Why do we yawn?

Yawning is a reflex—something that happens even if you don't think about it. When you yawn, you take a deep breath in through your wide-open mouth and then forcefully release it. There have been many theories as to why we yawn—perhaps because we are tired or bored or aren't getting enough oxygen in our lungs, but no one knows exactly why we do it.

We develop the reflex to yawn before we are even born. Doctors have taken ultrasound pictures of babies yawning while they are still in their mother's belly. People are not the only animals who yawn—reptiles, birds, and fish do it too! In humans, yawning can be contagious. If one student yawns in class, chances are other classmates will do it too. But we still don't know why—if it is the sight, sound, or smell of yawning that causes us to copy it.

What is your Adam's apple?

Your mouth is connected to your airway (trachea), which is connected to your lungs. The first part of your airway is called the larynx—it is made of cartilage, the same firm material that makes up your ears and the tip of your nose. Part of the cartilage sticks out in the front of your throat and is called the Adam's apple. The larynx is responsible for protecting you from choking, to make sure the food you eat goes down your swallowing tube (esophagus) and not your airway. The larynx is also important because it contains your vocal cords and allows you to speak.

The Adam's apple in men is more easily seen than in women because a man has a larger larynx than a woman, which provides a deeper voice. If you watch a man sip water, you can actually see his Adam's apple bob up and down with each sip, guiding the water away from his airway and into the esophagus on its way to the stomach.

Why do we get scabs?

When you get a cut, the first thing your body does is constrict or narrow the blood vessels in that area to prevent further bleeding. Proteins within the blood, like fibrin, form a protective layer over the cut called a scab. The medical term for scab is eschar. A scab serves several purposes. First, it protects the cut from getting contaminated by outside germs while it is healing. Second, it helps to seal off the blood vessels in the area to prevent bleeding. Last, it provides a protective roof or shelter for the skin cells under the scab to reach out to one another and join, thus healing the wound.

Because scabs are so important to wound healing, you should *not* pick them. It could delay wound healing or even make a scar worse, so leave that scab alone. The scab eventually will fall off when the skin beneath it is sealed properly. If the scab is in an area that moves a lot—like an elbow or a knee, you can cover it with a loose bandage to prevent it from getting dislodged before it's time to come off.

Why do some people have crooked teeth and others have naturally straight teeth?

Some people are tall, others are short. Some people have crooked teeth and others have straight. If your parents have or had crooked teeth, you probably will too.

The orthodontist is the dental specialist who deals with malocclusion of the mouth, which is abnormal positioning of the teeth or jaws. The most obvious reason that someone needs braces is because their teeth are not straight and it makes their smile look bad. But there are a whole bunch of other reasons why the orthodontist might recommend braces. If your teeth are crooked, it may affect your ability to bite and chew your food properly. Malocclusion can make it difficult for you to clean your teeth, and this can lead to cavities or gum disease. Sometimes crooked teeth can cause abnormal speech development or improper jaw growth.

Braces are brackets (usually made of metal) that are connected by a wire that is in the shape of an arch. By tightening this wire periodically, the teeth are gradually moved into proper position. For most people, this process usually takes eighteen to thirty months, after which they need to wear a retainer for six to twelve months to prevent the teeth from moving back to the wrong location. There are newer types of braces made of clear material that are not as noticeable and may be used in certain circumstances.

Why do you get goose bumps
when you are cold?

When you are outside on a winter day, the body has several ways to maintain its core temperature in the range of 98.6 degrees F. For example, when you shiver, you are making work for the body and increasing heat production. The fatty layer of tissue under the skin serves as an insulator to help prevent loss of body heat. The blood vessels near the surface of the skin narrow to prevent the warm blood from reaching the cold skin surface, which would cause the body's heat to escape. And when you are exposed to cold temperatures, you get goose bumps—the correct medical term is piloerection. In other words, the tiny muscles at the base of the skin hair shafts contract, causing the hair to stand up straight and producing little bumps on the skin. Furry animals get goose bumps too, but when the fur rises, it produces a layer of still, unmoving air above the skin, providing insulation. In humans, since we are not furry, getting goose bumps serves no known purpose.

You can also get goose bumps when you are afraid. When animals get goose bumps and raise their fur, they appear larger and can frighten off their enemies. One great example of an animal with this form of defense is the porcupine. When a porcupine gets threatened, it puffs out its quills and looks really fierce!

What is the black circle in your eye?

||||||||||||||||||||||||||||||||||||||||||||||||||||||||||||||||||||||||||||||||||||||||

t he black circle is called the pupil. It is actually a hole in the front of the eye that is covered by a clear window or lens. When the doctor looks inside your eye with a lighted ophthalmoscope, he or she can see the back of your eye, including the optic nerve. This nerve is actually part of your brain. Looking in your eye through the pupil is the only way to see the brain from the outside—looking up your nose or in your ears will not give you a view of the brain.

The purpose of the pupil is to allow light to enter your eye. When it is dark, the muscles around the pupil cause the hole to get bigger and allow more light into your eye so you can see. When it is bright, the pupil gets smaller to prevent too much light from entering the eye. The aperture on a camera works much the same way, to allow you to take pictures in a variety of lights and settings.

Why is blood red?

b lood is mostly made up of a watery fluid called serum. Within the serum are the three main types of cells: red blood cells, which carry oxygen; white blood cells, which help the body fight infection; and platelets, which help the body to form blood clots if you get a cut or scrape. There are other things in blood too, like sugar and clotting products, but 40 to 50 percent of blood is taken up by the red blood cells. These cells are tiny round pillows with a dimple in the center. They contain the hemoglobin molecule, which is responsible for carrying oxygen to the body. Each hemoglobin molecule is composed of four iron atoms, and it is the red hemoglobin that colors the blood red.

What is your ear made of?

Your outer ear (the part you can see) is made up of the auricle and the lobule. The auricle is the part of your ear that looks like a shell. It is designed to act as a collecting trumpet to "catch" sound waves. It is made of cartilage, the firm, flexible material that makes up the tip of your nose and is in your joints to help with movement. The lobule, also known as the earlobe, does not contain cartilage. It is made up of skin and fatty tissue, and it feels soft and fleshy. So when you have an earlobe pierced, you don't feel much pain.

Why does your blood look blue when it is in your body?

||||||||||||||||||||||||||||||||||||||||||||||||||||||||||||||||||||||||||||||||||||||||||

I f you look at the underside of your wrist, you can see several tiny veins and the blood in them looks blue from the outside. On the inside, however, blood is red. If you've ever had to have your blood drawn for a medical test, you can see that the blood that is collected in the tubes is red, not blue.

Blood is pumped out by the heart, carrying oxygen picked up from the lungs. This oxygenated blood is carried all over the body by the arteries—to your organs and muscles, everywhere. It is bright red. When the oxygen is released from the blood into the body, the deoxygenated blood travels back to the heart and lungs via the veins. This type of blood is dark red, not bright red, because it doesn't have much oxygen in it. Since many of your veins are very superficial, right under the surface of the skin, the dark red blood appears blue to you through the covering of the thin-walled veins and your skin. The bright red blood contained in the arteries is hidden from view by the thick walls of the arteries. You can feel the blood flowing through your arteries by locating the pulse in the underside of your wrist near the thumb side.

Why do I laugh when I think something is funny?

humans are not the only animals who have the ability to laugh. Smiling and laughing have been observed in non-human primate species during social play. This type of behavioral response serves as a signal to the group by spreading positive emotions, decreasing stress, and contributing to the cohesiveness of the group.

Humor-evoked laughter in humans can be divided into three stages. When listening to a joke, the first part of the humor is the punch line, an incongruous ending. Second, your mind begins to problem-solve in order to interpret this incongruity or surprise. Finally, the brain is able to appreciate these steps, which together form humor and evoke a response of laughter.

The neurotransmitter dopamine (a brain chemical) is responsible for allowing the brain to progress through the stages of humor. Dopamine allows us to feel good when we laugh. Some studies have demonstrated an improvement in health for chronically ill patients when they are exposed to funny stimuli. Thus the old adage "Laughter is the best medicine" probably has a note of truth in it.

Why do human beings cry?

IIIIIIIIIIIIIIIIIIIIIIIIIIIIIIIIIIIIIIIIIIIIIIIIIIIIIIIIIIIIIIIIIIIIIIIIIIIIIIIIIIIIIIII

ears are a combination of water, oil, and mucus secreted by glands and cells in the eyelids. The tears are collected by tiny holes (puncta) located in the inner corners of the top and bottom eyelids. If you look closely at your eyelids, you can see these two little holes. Once the tears are collected in the puncta, they drain through the lacrimal sac into the nasal passages (which is why your nose runs when you cry).

There are three types of tears: basal tears lubricate the eyeball and prevent it from drying up; when you get a piece of dirt in your eye, reflex tears help wash it away; and emotional tears result when the eyelid glands produce too many tears to be drained away by the puncta. The emotional tears, caused by strong feelings like anger, sadness, failure, exhaustion, fear, anxiety, loneliness, or joy, overflow out of the eye.

Women cry more than men, averaging about five crying episodes per month versus one per month for men. Women usually have the type of tears that overflow and run down their cheeks, whereas men's eyes usually well up with tears but do not produce enough to roll down their faces.

We don't know exactly why human beings cry, but it is thought that negative toxins or hormones are released in the tears. When you are finished crying, you feel better because these substances have been eliminated from the body. Crying may also stimulate brain chemicals called endorphins, which can naturally improve your mood and decrease pain.

What are moles?

t he medical term for a mole is a nevus. It is a benign (non-cancerous) collection of specialized pigmented cells called nevus cells that live in the skin. Up to 98 percent of all light-skinned people will have moles by adulthood, usually between ten and forty moles. If your parents have a lot of moles, you are more likely to have moles as well. Sun exposure also increases your likelihood of developing moles. The average nevus is about six to eight millimeters in diameter (less than the size of the end of a pencil eraser), symmetric with a regular outline, and round or oval shaped. Most are brownish-black in color with even pigmentation. Occasionally, nevi can contain irregular dysplastic cells. People with dysplastic nevi have a much higher likelihood of developing a melanoma skin cancer. These people should be checked by a dermatologist at regular intervals.

Why do we form scars?

|||||||||||||||||||||||||||||||||||||||||||||||||||||||||||||||||||||||||||||||||||||||||||||||||||||||

a scar is formed whenever the skin is injured. Scarring is divided into three phases. The first forty-eight to seventy-two hours after the injury is referred to as the inflammatory phase, when the body produces inflammatory cells and sends them to the wounded site. Second is the proliferation phase, which lasts between three and six weeks. Specialized cells called fibroblasts form a complicated framework to bridge the skin defect. A protein called collagen is formed to help bring the two sides of the wound together. Last, the scar maturation phase can take up to two years for the bridging to be complete and for the scar to achieve its final and permanent appearance.

Everyone scars differently, depending on the size, depth, and location of the wound, as well as individualized features such as age and skin color. Some people form fine white scars that are difficult to notice. Others form hypertrophic scars, which are raised, reddish-pink, and itchy. Finally, people can also form keloid scars, which are raised scars that continue growing beyond the boundary of the wound and can form a large lump of unsightly scar tissue.

If the skin is injured by trauma or surgery, a scar will form— there is no such thing as "scarless" surgery, although some procedures require smaller incisions than others. For people who have injured themselves in an accident, stitches may be required if the wound is greater than one-quarter inch deep or is located on the face. Although scar hydration with oils, petroleum-based ointments, or hydrating creams can bring relief from pain, itching, or tightness, they do not improve the appearance of the scar. Silicone gel sheets, which are available over the counter, have been clinically proven to minimize the appearance of scars, particularly for hypertrophic and keloid scars.

What do your tonsils do?

Y our tonsils are the two balls of tissue in the back of your throat (behind your tongue) that help you to fight infection because they contain cells that contribute to your immune system. If you open your mouth really wide and stick out your tongue while looking in the mirror, you can see them just above the base of your tongue. The tonsils are very small when you are born; they grow as you do, reaching their biggest size when you are between two and seven years old. After that, the tonsils shrink a little. These size changes make sense: when you are young you are exposed to lots of coughs, colds, and germs, and you need your tonsils to help fight off these bacteria and viruses.

Sometimes your tonsils can get infected and swollen—a condition called tonsillitis. This type of infection can go away on its own or may require antibiotics if the infection is caused by bacteria. Most tonsillitis occurs in children, but adults can get it as well. Sometimes you can get repeated bouts of tonsillitis, or your tonsils can be so big that they can block your breathing while you are asleep (sleep apnea)—if these situations occur, the doctor may recommend surgery to have your tonsils removed. This operation is called a tonsillectomy—it is a very old operation, which has been performed for more than three thousand years. You can live without your tonsils because your body has other ways to fight infection through its immune system, like lymph nodes and white blood cells.

How does food give you energy?

a calorie is a way to measure the energy that your body uses every day. A car needs gasoline to run properly, and the human body needs calories to make it work too. You get calories from the foods you eat. Different foods have different calorie counts—a banana has about 110 calories. An apple has 125 calories. Some cookies have 200 calories. A hamburger has about 250 calories, as does a small packet of fries from a fast-food restaurant. You can check the nutrition label on foods that you buy at the grocery store to determine how many calories are in each serving.

Your body needs calories in order to survive. Every day your body needs its fuel to allow your heart to beat, your brain to think, and your lungs to breathe the air. Your body also uses calories when you exercise. The more you exercise, the more calories you burn up.

How many calories should you eat every day? That number is different for every person based on, for example, your height and weight or whether you are a man or a woman. In general, for most school-age children, you need between 1,600 and 2,500 calories a day. An average adult needs around 2,000 to 2,500 calories per day. If you eat too many calories and do not get enough exercise, you will gain weight. If you eat too few calories for the amount of physical activity that you do, you will lose weight. If you eat the right amount of calories for how active you are, you will maintain your weight.

It's important to get your calories from a wide variety of foods. These include fruits, vegetables, whole grains, dairy, protein—like meat, chicken, fish, or beans—and a small amount of fat. Remember that when you eat a balanced diet, you stay healthier and feel better.

Why do you have a belly button?

||||||||||||||||||||||||||||||||||||||||||||||||||||||||||||||||||||||||||||||||||||||||||||||||||||

everyone has a belly button—the medical term is umbilicus. Before you were born, you were inside your mother's belly. You couldn't eat or breathe air. Your umbilical cord connected your belly button to your mother's body through an organ called the placenta. Your mother sent you nutrition and oxygen through the bloodstream via the umbilical cord and you sent back wastes, which your mother's body eliminated. As soon as you were born, you were able to breathe in air through your lungs and drink milk. The doctor cut your umbilical cord and tied it off near your abdomen. It formed a little stump, which, after about a week, dried up and fell off, leaving the scar of your belly button.

Why does your mouth make spit?

the medical term for spit is saliva. There are special glands in your mouth that produce the saliva. These glands are located under the jaw, under the tongue, and on the side of the jaw. Your body makes about one and a half liters of saliva every day—that's about forty-eight ounces or six cups of spit!

Saliva is about 99 percent water and 1 percent proteins and salts. Your saliva does some very important jobs for your mouth. It keeps your teeth, gums, and tongue moist to help the food go down. The proteins in the saliva break up the food that you eat to start the digestion process even before the food gets to your stomach. Without saliva, your teeth would be unprotected from decay and cavities. Finally, the protein in your saliva also helps you to fight germs that get into your mouth. Some people have certain diseases or conditions that cause them to make too little saliva—in these instances, they can have pain, tooth breakdown, and difficulty swallowing.

What's a concussion?

||||||||||||||||||||||||||||||||||||||||||||||||||||||||||||||||||||||||||||||||||||||||||||||||||||

a concussion is an injury to the head—usually from trauma, as when you are in a car accident or get hit by a baseball bat or other similar blow. Most of the time you lose consciousness, meaning you are asleep and don't respond when people call your name. This loss of consciousness can last from seconds to minutes or much longer. When people wake up from a concussion, they are often confused and don't remember what happened to them. They can also feel dizzy, have a headache, or have blurry eyesight.

When you have a head injury, doctors need to be sure that you don't have any bleeding around the brain. Special imaging tests, like a CT scan or MRI, can confirm whether an injury is a concussion or something more serious. Remember to always protect your brain with a helmet when you are doing activities that could potentially cause a head injury—like bike riding. When you wear your helmet, you reduce your chances of any type of brain injury.

Why does your heart make a "lub-dub" sound when it beats?

|||||||||||||||||||||||||||||||||||||||||||||||||||||||||||||||||||||||||||||||||||||

Your heart is divided into four spaces or chambers. The top of the heart contains the two chambers called atria, and the bottom consists of the two chambers called ventricles. The heart is also divided into the right and left sides. The right side has the right atrium and the right ventricle; the left side has the left atrium and the left ventricle. Blood flows to the heart from the body and into the right atrium. It passes through a connection or valve into the right ventricle and then flows past another valve into the lungs to receive oxygen. Afterward, the blood flows into the left atrium, through a valve into the left ventricle, and out through another valve to the rest of the body. The "lub" sound is from the closing of the valves between the atria and the ventricles. The "dub" sound is from the closing of the valves leading out of the ventricles.

By listening to your heart with a stethoscope, the doctor can determine whether your heart sounds, labeled sound #1, or S1 ("lub"), and sound #2, or S2 ("dub"), are normal and whether your heart is functioning properly. People who have heart problems will have extra heart sounds, labeled S3 and S4. If S1 and S2 do not sound normal, there may be a problem with one of the heart valves. This abnormal sound is called a heart murmur.

What is a charley horse?

a charley horse is a muscle cramp, often in the calf (back of your lower leg). The pain is sharp—almost like being kicked by a horse (which is a possible theory of the name's origin). A charley horse occurs when a muscle contracts and does not relax. People describe it as a sudden muscular pain or knot. It usually happens during exercise, like running or swimming/kicking, but can also occur at rest for no reason at all. People who exercise and don't drink enough water become dehydrated and are more likely to develop muscle cramps.

The treatment for a charley horse is to stop exercising and gently stretch the muscle. It may also help to massage the area and to try to knead out the knot. Some people get relief from a warm compress or steamy shower. Most charley horses go away on their own after a few minutes and do not require a trip to the doctor's office.

What is a cell?

a cell is a building block for your body—like bricks are for a house. But cells are different from bricks—they are not just used to construct your body. Cells are made up of about 90 percent water. Each cell is like a miniature factory that contains an energy source as well as special information (genes) that are unique to every living being. Some animals are made up of only one cell, like an amoeba. When you eat an egg for breakfast, you are eating one cell! Human beings are made up of trillions of cells that are so tiny they can only be seen with a microscope.

Human beings have six different kinds of cells in their bodies—blood cells, nerve cells called neurons, muscle, fat, bone, and skin cells. Some cells live for only a short period of time, like skin cells, which can die as quickly as every twenty-four hours. Dry, flaky skin is actually dead skin cells being sloughed off your body. But skin cells can divide and form new cells, so you never run out of skin. Skin cells can only make new skin cells. They can't make other types of cells, like bone or muscle. Other cells, like some of your nerve cells, live as long as you do and can't divide to form new cells. That is the reason why when someone is paralyzed from an accident (meaning the nerve cells in their spinal cord have been damaged) they can't grow new nerve cells to fix the problem. Scientists are working hard to develop ways to coax nerves cells to divide and regenerate so many neurologic diseases will be cured.

Why is a bruise black and blue?

||||||||||||||||||||||||||||||||||||||||||||||||||||||||||||||||||||||||||||||||||||||||||||||||

I f you trip while jogging and bang your leg, you will get a bruise. The medical term is hematoma. Inside your leg, small blood vessels burst from the trauma and red blood cells leak out into the fatty tissue just below the skin surface. The red blood cells are made up of hemoglobin (the stuff that carries oxygen), which is red, so immediately after you fall, the bruise can look red, red-purple, red-blue, or even red-black. As the hemoglobin breaks down, it turns into a substance called biliverdin, which is green, and then bilirubin, which is yellow.

If you have a bruise, you can know when it will heal up by looking at the colors. If it is purple-red, it is a fresh bruise, just beginning to heal. If it is yellow-green, it's almost done healing and you should feel better soon.

Why do some people have freckles?

Your skin contains special cells called melanocytes that carry the pigment called melanin. This melanin protects your skin from sun damage and potentially from developing skin cancer. If you have a lot of melanin in your skin, you have dark skin. If you have just a little melanin, then your skin is fair.

Some people, especially fair-skinned people, have melanin that is not evenly distributed—meaning there are tiny pockets of melanin-containing cells in the skin. These small spots show up as freckles, which are brownish colored, flat, and about the size of a pinhead. Sometimes you may have a lot of freckles that all blend together, which makes it look like you have very large freckles. The more your skin is exposed to the sun, the more freckled it becomes. Freckles are found most commonly on sun-exposed parts of your body—like your nose, cheeks, and forearms.

Why do some people have dimples?

|||||||||||||||||||||||||||||||||||||||||||||||||||||||||||||||||||||||||||||||||||||||||||||||||||||||||

dimples are those adorable little indentations on your cheeks. They are caused by puckering of the muscle under the skin when you smile. Your cheek muscle is called the buccinator. If you want to find this muscle, just pretend you are blowing into a trumpet—you'll see the muscle in action! Dimples are present when you are born, but they can change as you get older or gain or lose weight. They are equally common in women and men.

Dimples are hereditary—if one or both of your parents have dimples, then you may get them too. I don't have dimples, but Andrea and Julia's father has a dimple on his left cheek only. Andrea and Julia did not inherit his dimple. Some people can fake dimples by sucking in and biting the inside of their cheek with their teeth. Others even go to the extreme of having dimples made for them—a surgeon can actually put a stitch inside the mouth and attach the muscle to the undersurface of the skin (if you want dimples that badly!).

What are genes?

||||||||||||||||||||||||||||||||||||||||||||||||||||||||||||||||||||||||||||||||||||||||

each person is different. You may be tall or short, blond or brunette, brown- or blue-eyed. Have you ever wondered why everyone is unique? It's because each person has cells that contain a special set of instructions for that human being. The instructions, or DNA (deoxyribonucleic acid), tell your body's cells how they should develop and grow. DNA is divided into smaller sections called genes. Human beings have about 35,000 genes, which determine everything, from how we look outside to how our body functions inside.

The location of each of these genes along our DNA is called the human genome. It is a giant map showing where each gene lives in relation to other genes. It's sort of like an enormous CD with individual songs in a certain order. In April 2003, scientists from all over the world finished mapping each of these genes to form a complete map of the human genome. You can celebrate National DNA Day with these scientists every April and remember how important it is to have such detailed information about our hereditary makeup.

Researchers are using the human genome map to help develop ways to cure diseases like cancer. Scientists also use DNA for criminal investigations. If a robber leaves a drop of blood or a single hair behind, the police can use the DNA in the blood or the hair to match it to the thief. If a criminal writes a letter to the police, they can trace the DNA in the saliva used to seal the envelope in order to catch the crook.

What happens when you break a bone?

‖‖

b roken bones or fractures are relatively common, especially in children. Signs that someone may have a broken bone after an injury include pain, swelling, and bruising. Sometimes you can see that a bone is broken just by looking at the arm and seeing that it is no longer straight or it is pointing in an unnatural position. An X-ray can tell for sure if it is broken.

Even as an adult, your bones are constantly growing and changing, even though you don't know it. Some bone cells are designed to change the structure of the bone and other bone cells make your bones grow stronger. This constant remodeling is taking place all the time in children and adults. If you break a bone, the bone cells can heal the fracture by making new bone growth. The doctor will first put a cast on you to make sure the two broken ends of the bones match up. That way, when new bone forms it will be a smooth and straight bone. If you don't line up the two bones, the new bone can be lumpy and too short or even too long, so wearing the cast is important. Most fractures are simple breaks, so they heal quickly over several weeks.

The Weird,
the Ugly, and the
Downright Gross

It is amazing how the human body works—how muscles and bones coordinate when you run, how your brain is capable of storing memories, and how your senses allow you to interact with the environment. But not everything about your body is elegant. There are some parts of your body that are just weird, ugly, and downright gross. This part is all about nose hair, bad breath, yellow teeth, and burping. We try to answer every repulsive body question that you can think of, from the top of your head down to your feet—including why some people have six toes!

Why don't your eyelashes grow?

||||||||||||||||||||||||||||||||||||||||||||||||||||||||||||||||||||||||||||||||||||||||||||||||||||

Your eyelashes do grow, but not very fast.

Every hair on your body goes through a cycle—first the hair grows, then it transitions to a resting phase when it stops growing, and eventually it falls out. The hair on your head has a long cycle that lasts for many years—scalp hair goes through its growth cycle for several years, followed by a two-to-three-week transition phase and a resting period of three to four months before falling out.

On the other hand, eyelash hair goes through a short cycle of growth (two to three months), then a transition phase (two to three weeks), followed by a resting period of two to three months before the eyelash falls out. The cycle then begins again with new lashes. Because your eyelashes live only several months—compared with your scalp hair, which grows for several years—your eyelashes never grow enough for you to notice. Finally, because individual hairs start and end their growth cycles at different times, all of your hair never falls out at once!

Why doesn't it hurt when you get a haircut?

Your hair is made up of two main parts—the hair follicle and the hair shaft. The follicle is the root that sits inside the scalp and contains the part of the hair that is alive and can feel pain. The hair shaft is the part of your hair outside the scalp that you can see. It is made up of protein called keratin—but this protein does not contain living cells. Therefore, when you cut the keratin protein it doesn't hurt. But you do feel pain if your hair is yanked out of the scalp, because the hair is being pulled out of its root, the living part of the hair.

Why do some people have "innie" belly buttons and some people have "outies"?

everyone has a belly button (the medical term is umbilicus), because the belly button is where the umbilical cord connects the child to its mother during fetal development. Most people have "innie" belly buttons, but some people are born with a hole in the belly wall that allows their intestines to poke through the belly button, creating a hernia or "outie." This hole is caused by a failure of the belly wall to seal around the space that was taken up by the umbilical cord, which is tied off at birth. Most of these umbilical hernias close by themselves as a child grows, but occasionally they are very large and require a small outpatient operation to close the hole. If needed, this surgery is performed early in life.

Why do you see fireworks when you close your eyes and press on them?

When your eyes are open you can see people and objects in all colors because light stimulates the back of your eye so that you can see various images. But you can also trick your eyes into "seeing" when your eyes are closed by pressing on your eyelids to stimulate them. This manual pressure excites the back of the eye (retina) to produce the perception of flashes or spots of light that look like fireworks. These light flashes are called phosphenes.

Why do we get wrinkles as we get older?

a s we get older, our skin gets thinner and thinner, and thus loose and wrinkly. The skin contains a protein called elastin, which is like a rubber band. The older we get, the less stretchy the elastic becomes, and the skin loosens up—sort of like a pair of knee socks that don't have any elastic left and keep falling down around your ankles.

Some people are more likely to develop wrinkles—if your mother or your father has a lot of wrinkles, you may have a lot of them too. Although everyone gets some wrinkles when they age, there are things you can do to minimize them. First of all, wear sunscreen every time you go out in the sun. When you get a tan or a sunburn, you are damaging your skin, making more wrinkles later. Also, smoking cigarettes causes the skin to lose its ability to stretch—if the skin can't stretch, it gets saggy and wrinkled. Last, people who gain a lot of weight and lose a lot of weight stretch their skin, causing even more creases.

Why do we burp?

there are many names for burping—belching, aerophagia, and eructation—but they all mean the same thing. When you swallow excessive air, it gets into the top part of your digestive tract (esophagus and stomach). As this air comes up and out your mouth (instead of going farther down your digestive tract), it is called burping.

Any repetitive mouth behaviors can lead to swallowing extra air—like gum chewing, smoking, drinking through a straw, talking a lot, or eating very quickly. After you've eaten a big meal, you may have swallowed a lot of air and feel like burping to relieve the pressure. The old adage "It's better to burp and bear the shame than not to burp and bear the pain" still applies. Many people feel much better after they burp. If you are burping a lot, try to avoid sodas and other carbonated beverages, eat slowly, and don't chew gum. If you don't swallow excess air, you won't feel the need to burp.

Why is your poop brown?

When you eat food, it passes through your stomach and intestines—it is the gut's role to absorb nutrients while the food is traveling through it. Your liver (an organ inside your abdomen) secretes a substance called bile, which helps the intestines digest the food. Bile is made up of several different components, one of which, bilirubin, is formed from the breakdown of red blood cells. As the bilirubin passes through the intestines, bacteria interact with it to form a brown pigment that makes your stool brown. If the liver is blocked by certain types of diseases, the bilirubin doesn't get into the intestines and the brown pigment is not made—leaving the stools white (the medical term is acholic).

Why do we get gray hair?

Whenever Andrea and Julia ask me why I have gray hair, I always say it's from worrying about them! Although stress can contribute to gray hair, aging is the real cause. People get gray hair when the melanin-producing cells die. Melanin is the pigment that causes the hair to have color. As you age, these cells die and the hair loses its color and becomes almost see-through, appearing as white, gray, or silver. Some people can start turning gray when they are teenagers, but most people don't start getting gray until they are in their thirties and forties. Some people turn gray very quickly, in just a few years, but most people take more than ten years before all of their hair turns gray, and some people never get gray at all. If you know when your parents started going gray and how much gray they have, you may get some idea of how old you will be when you start to turn gray.

Why do we sometimes feel nauseated while in a car or on a bus?

||||||||||||||||||||||||||||||||||||||||||||||||||||||||||||||||||||||||||||||||||||||||||||||||||||||||||||||||

m ost people will experience motion sickness at some point in their lives. It is more common in children, and some will outgrow it by adulthood. Feelings of dizziness, nausea, and vomiting occur when your sense of equilibrium (balance) doesn't match what your other senses are reporting to you. For example, if you are reading a book while taking a drive in the car, your inner ears and brain sense motion, but your eyes do not because you are busy reading.

If you start to feel motion sickness, you should look out of the car window so that your sight and touch feel the motion of the car in coordination with the balance receptors of your inner ears. If you know you have a problem with motion sickness, try to sit by the wing of the airplane, where the ride is less turbulent. It also helps to sit at the front of the bus, which is a less bumpy ride. Some medicines may prevent motion sickness if you take them before you get in the car or on the ship or plane. These medicines can either be in pills or a patch that sticks to your skin.

What are lice?

||||||||||||||||||||||||||||||||||||||||||||||||||||||||||||||||||||||||||||||||||||||||||||||||||||||||||||||||||||||||||||||||||||

t he medical term for the louse is *Pediculus humanus,* and the types that infest the head and the body, respectively, are termed *capitis* and *corporis.* Lice can live for only a week without a human host. They are passed from person to person by shared clothing and hair items like brushes, ponytail holders, and headbands, as well as direct contact. Each louse is only two to three millimeters in size (the size of two or three pinheads) and looks like a small flat sesame seed. Once on the human being, each female louse can lay five or six eggs a day, which are attached to individual hairs on your head. Then the eggs (called nits) hatch in eight to ten days and become adult lice in two weeks. The lice have two meals per day—they eat by puncturing the scalp and eating your blood! Eventually, these puncture sites become inflamed, red, and itchy— that's when you know you have a lice problem.

The best way to treat lice is to never get them in the first place— don't share hairbrushes, combs, hats, or pillows that would allow the lice to transfer themselves from one person to another. Kids are more likely to get lice because they often share these items at school or slumber parties. If you know someone who has lice, avoid direct contact with them. If you do get lice, you will need to use a special shampoo. Then someone will need to comb through your hair thoroughly—picking out both lice and nits. Any clothing or bedding that has come into contact with the lice will need to be washed in hot water and dried in a hot dryer. If some items cannot be washed (like large pillows), they will need to be placed in a plastic bag for about two weeks to make sure any lice will die—remember, the lice cannot survive without a human being. Many people think that if you get lice it's because you are unclean or have dirty hair and clothes. That's not true—anyone, no matter how clean and hygienic, can get lice from another person. So you don't have to be embarrassed, but you do need to get rid of the lice completely.

Why do we have nightmares?

|||||||||||||||||||||||||||||||||||||||||||||||||||||||||||||||||||||||||||||||||||||||||||||||||||||||

h ave you ever looked at people when they were asleep and seen their eyes moving quickly and their arms or legs twitching? This kind of sleep is called rapid eye movement (REM) sleep, and it makes up about 20 percent of our sleep. The other 80 percent is called non-REM sleep. Non-REM sleep has four distinct parts, ranging from drowsiness to deep sleep. After all four parts have been completed, you pass into REM sleep, which is closer to wakefulness and the period in which you dream.

Nightmares are really scary dreams that are frightening and very detailed. You can often remember them exactly, and still be scared even when you are fully awake. Some common nightmare themes include being chased by bad people or monsters or being lost and unable to find your way home. Most people do not call out or scream, but they do wake up quickly. These nightmares occur during REM sleep, so they occur late in sleep, not right away when you are still going through the non-REM sleep stages. They are usually infrequent and do not recur.

Nightmares are more common in children and no one understands completely why we have them, but they are a part of dreaming. People who have recurrent nightmares may have an underlying psychiatric disorder, such as post-traumatic stress disorder.

Why do toenails grow slower than fingernails?

Your nails grow from the white half-moons that are at their base, also called the matrix. This matrix forms keratin, a protein, which then makes up the nail plate. Because the keratin is not made up of living cells, it doesn't hurt when you cut your nails. Keratin is the same protein that makes up horse hooves and animal claws. If you catch your finger in the car door, it definitely hurts because the matrix part of the nail contains living cells with nerves. If the car door deforms the half-moon matrix, the new nail will grow out with a bump or ridge.

Your fingernails grow at the rate of about one millimeter or one pinhead every ten days, but toenails grow much more slowly. One theory for this difference is that nail cells are stimulated to grow by daily wear and tear—for example, if you bite your nails, they grow faster. Since your fingers go through more wear and tear than your toes, the fingernails grow much faster than the toenails. As you age, the rate of nail growth slows and you don't have to cut them as often.

Why do people get pimples?

the medical term for pimples is acne—bumps on the skin that can be red or swellings topped with white or black dots. Acne is most common on the face and chest. Every hair on your face and chest lives in a tiny tube (called a hair follicle) that is moistened by an oil-producing gland. Usually this tiny amount of oil makes its way up the hair follicle and onto the surface of the skin to lubricate it. Sometimes, however, the opening of the hair follicle (pore) is blocked and the oil becomes trapped and collects under the skin surface, making a small bulge or pimple. Bacteria can get trapped inside the hair follicle and cause irritation, forming a pimple that way.

Acne is more common in teenagers as their bodies make developmental hormones, but pimples can occur at any age. If you get pimples, make sure to keep your face clean and try to avoid touching or popping the acne—this can leave scars or cause a bad infection. There are many over-the-counter creams and lotions to help with acne, and a doctor can also provide prescription-strength medicine.

Why do we have boogers?

the medical term for boogers is mucus. We form mucus in the nose to prevent dust or dirt from entering the nose and being breathed into the lungs. Usually the mucus is clear to white-colored, but if it turns yellow or green, you may have an infection—maybe a cold or ear/sinus infection. You can use a tissue to be polite and clear away the boogers, although some just evaporate and others are swallowed after they drip from the nose into the back of your throat (postnasal drip). Picking your nose can lead to a nosebleed or infection, so always use a tissue!

Why do people have hair in their nose?

You have hair in your nose to filter out dirt and to prevent dust from being inhaled down to your lungs. Your nose hairs have a name—vibrissae—which comes from a medieval Latin word, *vibrare*, meaning "to quiver." As you age, your body hair changes because of changing hormone levels—for example, the hair on your head and body usually becomes thinner and finer. Other hair, like your nose hair, tends to become coarser and thicker. Because men are hairier than women in general, many old men have a lot of nose and ear hair.

Why do farts smell?

the medical term for farts is flatulence; however, there has been a wide range of terms, from "silent but deadly" to "downwind." On average, people fart about thirteen times a day. How noisy these farts are depends on how fast the gas is expelled from the behind and how tight the sphincter muscle is. Flatulence is gas swallowed by the mouth or produced by the body. Ninety-nine percent of these gases (oxygen, nitrogen, hydrogen, and carbon dioxide) are odorless and colorless, but it's the less than 1 percent part that makes your farts smell so bad—hydrogen sulfide, ammonia, and methane.

The food you eat affects how much you fart and how smelly they are. Gas-producing foods like beans, vegetables (such as broccoli), prunes, and fried foods increase the number of farts you make. Foods like meat and eggs can make your farts smell bad.

Why do you get bad breath?

||||||||||||||||||||||||||||||||||||||||||||||||||||||||||||||||||||||||||||||||||||||||||||||||||||||||

b ad breath, or halitosis, is usually caused by the bacteria that live in your spit (saliva). These bacteria break up bits of food stuck in your teeth and gums, and it is this rotting food that gives off chemicals (hydrogen sulfide and methylmercaptan) that smell bad. The halitosis is found on your tongue (especially the back part) and in between the teeth and gums. When you are sick with a cold, sinusitis, or tonsillitis, the infection makes your breath smell worse. Certain foods like garlic and onions can also give you bad breath or make your bad breath much worse.

Most people who have bad breath don't even know they have it, because your nose gets used to the smell after a while. One of the best ways to check for bad breath is to lick the inside of your wrist and let it dry for a couple of seconds. Then smell your wrist— you will know if your breath is bad! If you do have halitosis, you need to brush and floss regularly. Don't forget to brush your tongue, especially the back of it. You can also buy a tongue scraper, which can brush away the nasty coating that builds up on your tongue and contributes to bad breath. See your dentist at regular intervals and avoid smelly foods. Your breath is always worse in the morning because that's when your mouth is dried out, so re-member to drink enough water during the day to avoid dry mouth and bad breath.

What are warts?

‖‖

Warts are an infection by the human papilloma virus (HPV), which causes the skin to make too many skin cells. When you have warts on your feet, they are called plantar warts, or verrucae. It's easy to identify a wart—they look like small bumps, usually the same color as your skin. They feel rough when you touch them and look like a miniature version of a head of cauliflower. Plantar warts usually grow on the part of your foot that supports your weight, like the ball of your foot or the heel. They can cause pain and itching.

Warts are contagious. They can be spread from person to person by direct contact. The virus likes warm, moist environments like the bottom of the shower or tile floors in the pool changing rooms. Always wear flip-flops or some form of shoe cover when you are in public swimming areas or locker rooms. Your doctor can remove warts by using acid to burn them away or liquid nitrogen to freeze them away (it doesn't hurt, even though it sounds like it would). Never pick or scrape at the warts yourself, because any pieces you pick off still contain the virus and can spread to other parts of your body—your hands or other foot, for instance. Warts are not dangerous, but they can be embarrassing, so most people who have them usually seek medical care.

Can someone really have six toes on one foot?

||||||||||||||||||||||||||||||||||||||||||||||||||||||||||||||||||||||||||||||||||||||||||||||||||||

ost people are born with five fingers on each hand and five toes on each foot. But some babies are born with more than five. This condition is called polydactylism. It usually manifests as an extra little (or pinkie) finger, but people can have six, seven, or more fingers and toes on each side.

The extra fingers usually occur on the thumb or pinkie sides of the hands, and the extra toes are usually found growing off of the big toe or pinkie toe. Occasionally an extra finger or toe will grow from the middle of the hand or foot. Having an extra digit can run in families—so if your mother or father has or had one, you are more likely to be born with one. Rarely these extra fingers and toes can be a marker for an underlying chromosomal abnormality or other type of disease, so the pediatrician will want to get information on all family members to see if there is a serious problem.

There are two kinds of extra fingers and toes. In the first kind, the finger or toe looks like a real additional digit. It contains bones and joints and may have a fingernail as well. This is called a well-formed digit. Most people will opt for surgery to remove this extra digit to make it easier to be fitted for shoes and to walk, or because it is embarrassing to have an extra finger. The second kind of extra digit is not well formed. It is usually a little skin sac of fat, shaped like a teardrop, sticking off of the hand or foot. Because there is no bone in this kind of basic digit, the doctor can take a strong silk thread and tie off the extra finger/toe at the base where it starts. A tight knot is left in place for several days, and eventually the nubbin of finger or toe dies, turns black, and falls off, leaving hardly any scar. This procedure, which is painless, can be done in the hospital's newborn nursery.

Why don't human beings have tails?

human embryos (babies that are first developing inside their mother) do have tails—up to the eighth week of gestation, and then they disappear during further development. You can feel what is left of your tail—the tailbone or coccyx—by sitting cross-legged on the floor. In this position, you can feel the pointed, triangle-shaped bone at the bottom of your spine just before the crack in your buttocks. The coccyx got its name from the Greek word *kokkyx*, which means "cuckoo bird," because the bone is shaped like the beak of a cuckoo.

No one knows exactly why human beings no longer have tails like monkeys do, but one idea is that monkeys need their tails for balance and grasping as they swing from tree to tree. Since human beings walk upright on two feet, having a tail behind them would throw them off balance and make them fall backward. Thus, human beings have no need for tails.

Very rarely, a human baby is born with a small tail. This phenomenon was first described in the late 1800s and there have been fewer than one hundred cases reported in the medical literature since that time. Sometimes the tail is just made up of a skin sac filled with fatty tissue, which can easily be removed by a surgeon. But sometimes the tail can have a connection with the nerves of the spinal cord and requires delicate surgery by a neurosurgeon to remove it and close the connection safely.

Why do your teeth turn yellow?

Your teeth can turn not only yellow, but brown, gray, and black too! There are two reasons why your teeth become discolored. The first is an extrinsic reason—meaning you stain the outer surface of your teeth by drinking coffee, tea, or red wine, eating carrots, or smoking cigarettes. The second reason is intrinsic, meaning there is something wrong inside the tooth. Examples of intrinsic staining include exposure to certain antibiotics such as tetracycline during childhood tooth development, which leaves dark spots, trauma to the tooth, or thinning outer enamel of the tooth as you age, which makes the tooth appear darker.

The best way to get rid of external tooth stains is to brush twice a day and to floss regularly. Visit your dentist twice a year to get professional polishing and keep those teeth pearly white. One year, my daughter Julia received an anonymous Valentine card from a classmate saying that her teeth were as white as pearl diamonds— now that's white!

If your stains are really bad, you may need to talk to the dentist about tooth whitening—either strips that you use at home containing hydrogen peroxide or professional bleaching in the dentist's office. If the dentist says it is okay to use the at-home kits, be sure to follow the directions carefully. Too much bleaching can cause puffy gums, tooth sensitivity, and even weakening of the teeth by thinning the enamel.

Why do some people have bad body odor?

P erspiration or sweat comes from sweat glands. You have about three million sweat glands spread throughout your body. The majority of these glands are called eccrine glands. They are located in the skin, especially in the armpits, palms of your hands, and soles of your feet. The fluid that they secrete is to help you stay cool, but it does not smell bad. The second type of sweat glands are called the apocrine glands. These glands are concentrated in the armpits and the groin and become active during puberty. When bacteria on the skin breaks down the fluid secreted from apocrine sweat glands, it smells bad and is called body odor, or B.O. for short.

Many different things affect your body's odor, including the foods that you eat. If you like strong, spicy foods like garlic, cumin, and curry, your body odor will be worse than if you ate only a bland diet. Perfume can cover up body odor a little bit, but proper hygiene—for instance, using soap and deodorant and shaving your armpits—will improve the situation.

Each individual person has their own unique body odor, but identical twins smell very similar. How do we know this information? Scientific studies have been done in which each twin was asked to sleep with a piece of cotton gauze in his or her armpit. Then, in the morning, that gauze pad was turned in to the study investigators. Volunteer "sniffers" then smelled all the gauze pads and couldn't tell the difference between one twin and another. I wouldn't want to volunteer for that job—would you?

Why do your fingers and toes look like prunes when you come out of the pool?

Y our skin is the largest organ in your body. For an average man, the skin weighs approximately twenty pounds. That's heavy! The skin is divided into three layers: the inner layer, called the subcutaneous tissue, the middle layer, named the dermis, and the outer layer, which is the epidermis. Think of the old schoolyard joke, "Your epidermis is showing!"

The epidermis is approximately the same thickness in all parts of your body except for the palms of your hand and the soles of your feet. Here, the epidermis is much thicker than the rest of your body—usually about five times thicker. The outermost aspect of the epidermis contains old, dead skin cells, filled with a protein called keratin. When you take a long, hot bath or swim in a pool, the keratin soaks up the water and swells the epidermis. But because the epidermis is attached to the dermis, it cannot puff up completely; rather it folds in on itself, forming wrinkles or "pruny" fingers and toes. Once you come out of the water and the extra fluid evaporates, the skin looks normal again.

Why are some people able to wiggle their ears?

||||||||||||||||||||||||||||||||||||||||||||||||||||||||||||||||||||||||||||||||||||||||||||||||||||||

C ats and dogs can perk up their ears when they hear a sound, so why can't humans? Your ears do move when you are talking, yawning, and swallowing, but it's hard to move them without moving the rest of your face. The ability to wiggle your ears may be genetic and run in your family. My daughter Andrea is able to wiggle her ears (she's better at her left than her right), and her grandfather and great-uncle can wiggle theirs as well.

The muscles that allow you to wiggle your ears are behind and on top of your ears—called the auricularis posterior and auricularis superior muscles ("auricular" refers to the ears). You can try to train yourself to move these muscles voluntarily by pressing on them with your fingers to try to find them. Then open and close your mouth to feel how the muscles contract as you move your jaw. Start with one side and then the other—it's very hard to wiggle both ears at the same time. More men than women can wiggle their ears. With a lot of practice and patience, you may be able to figure out how to wiggle yours too.

Why is pee yellow?

Your urine is normally an amber-yellow color because of a pigment called urochrome, which was first discovered in the 1800s. If you don't drink enough fluid and are dehydrated, your urine becomes concentrated and turns a darker shade of yellow. If you drink a lot of water, the urochrome is diluted and your urine turns a very pale shade of yellow.

Urine doesn't have to be just yellow—it can range in color from clear to almost black (caused by a rare, inherited genetic disorder). It can be red or reddish brown, usually from blood in the urine (from a kidney stone, for example) but sometimes from medication or foods. If you eat too many breath mints, your urine can turn bluish, or if you eat a lot of asparagus, your urine looks green. Other people, because of illness, can have milky white, orange, or even purple urine. But for most people, most of the time, it's just plain old yellow.

What are bedbugs?

bedbugs are small (one-quarter inch) oval brownish insects that live by feeding on the blood of warm-blooded hosts. The scientific name for the most common type of bedbug is *Cimex lectularius*. Bedbugs live in beds, sheets, clothes, luggage, and furniture. They are found in people's homes as well as in hotel rooms, dormitories, cruise ships, and prisons.

People who are bitten by bedbugs will develop itchy red welts on their skin. Careful inspection of the bed linen will often reveal these tiny insects. Bedbugs do not transmit diseases, but they are annoying. Over-the-counter creams will take care of the itchy bites, but careful hot water laundering, cleaning, and vacuuming are essential in getting rid of the bedbugs permanently.

Why do some women have mustaches?

Women and men look different because they have different hormones. Male hormones, like androgen and testosterone, make men have beards and mustaches, and hair on their chests and abdomens as well as their backs and thighs. The male hormones are responsible for men having deep voices as well. Women have a light growth of hair all over their faces—usually it is a hard-to-see downy coating. But if women have too much of the male hormones, they can develop thicker, darker hair in places that they don't want—a mustache, for example.

Many medical conditions can cause this hormone imbalance. Smoking can also contribute to unwanted hair growth in women as well as being obese and taking certain medications.

No woman wants to have a mustache, and it can be a very embarrassing problem. Although there are some ways to get rid of unwanted facial hair, like shaving, depilatory creams, waxing, electrolysis, or laser treatments, these methods can be painful and expensive. Sometimes the hair can continue to grow back.

Why do some men have boobs?

men have nipples, but they're not supposed to grow breasts like women do. Sometimes, if men are overweight or obese, their chests can become fatty and they look like they have breasts, but they don't have much breast tissue. However, some men can actually grow breast tissue and look like they have female breasts—this condition is called gynecomastia. The breast growth is usually caused by medications or illness. Many men in this situation opt to have plastic surgery to remove the breast tissue growth so they are not embarrassed by their chest. This problem can also occur in teenage boys who are going through puberty—this type usually goes away by itself after several months.

Why do some people have fat
that shakes like Jell-O?

the Jell-O look or orange-peel appearance is called cellulite. It happens when the fat immediately beneath the skin puckers and dimples, giving the skin a lumpy appearance. The medical term for cellulite is dermopanniculosis deformans, and it is more common in women than in men because of female hormones. It is found most often on the hips, thighs, and behind. It may have a genetic component, meaning if your mother has cellulite, you may develop it too. There is no cure for cellulite, although there are many creams and lotions that claim to improve the appearance of the dimpling. There is no evidence that any of these work, so the best you can do is avoid being overweight, don't gain or lose weight rapidly, drink plenty of water, eat a healthy diet, and don't smoke (which decreases the elasticity of the skin, and thus exacerbates the problem).

Why do people vomit?

everyone knows how nasty it is to throw up, toss your cookies, hurl, or various other terms including the medical name—emesis. Throwing up is divided into three parts. First is nausea, the unpleasant sensation you experience that signals you may throw up. Your face may become flushed and red. Your heart rate goes up and your mouth makes extra saliva. Next comes the actual vomiting part. The muscles in your abdomen tighten up and the entry to your stomach opens up and forces anything in your stomach up and out your mouth. Yuck. The last part is retching—that's when your chest and abdominal muscles spasm without actually causing you to vomit.

A vomiting center in your brain is responsible for stimulating you to throw up. Some people may vomit because of illness or medications. Other people vomit because of something they taste (I gag whenever I'm forced to eat eggplant) or something they smell (Andrea and Julia retch when they have to open up a new container of dog food because of the odor) or from pain or distress (if you break a bone, for instance, or get really, really scared). You can also vomit if your gag reflex is stimulated. This reflex is activated when something touches the back of your throat—like if you stick your toothbrush too far inside your mouth or when the doctor sticks in a cotton swab to do a culture for strep throat.

Overall, vomiting is nasty business. But the good news is that once you do vomit, most people feel better and the nausea subsides immediately!

What does it mean to be double-jointed?

the medical term for being double-jointed is hypermobility, and it means looseness of the joints. About 10 percent of people are hypermobile. Sometimes double-jointedness can be associated with serious medical illness, but most of the time it is found without any additional problems. Younger people are generally more flexible than older people. If you want to see if you are double-jointed, see if you can do some of these moves:

1. Can you put your palms flat on the floor without bending your knees?
2. Can you bend your thumb to touch your wrist?
3. With your hand held out flat (palm facing the ground), can you pull up your pinky to a 90-degree angle, so that it points to the sky?

Hypermobility is more common the younger you are, and goes away as you age. It is more common in women, and if someone in your family is hypermobile, you have a higher chance of being double-jointed as well. If you are double-jointed, it can be fun to entertain your friends with contortions and twists, but always remember to be careful. Being hypermobile can lead to long-term wear and tear of your joints, resulting in arthritis, dislocation, or even a bone fracture.

Why do people bite their nails?

biting your nails is an example of a habit—a pattern of behavior that occurs without thinking about it. Nail biting is called onychophagia, and it is extremely common—almost half of all teens bite their nails at some point. It is much less common in adulthood.

Biting your nails can lead to many complications. Some are minor, like scarred or unattractive nails, but some are more serious, like skin infections or having a nail fall off and not grow back! There are many ways to discourage nail biting. One of the most common is to paint the fingernails with a bitter polish, so whenever you try to bite them, they taste awful. Behavioral therapy, like positive reinforcement, is sometimes helpful. You can try making a fist whenever you feel like biting your nails. It is impossible to bite your nails when your hand is in a fist. Eventually you will learn to reverse your nail-biting habit and replace it with making the fist. With time, you will be able to stop making the fist and the habit will be gone.

Why do tears taste salty?

Your body is made up of about 70 percent water. There is water in your blood and water in your cells. This water is mixed with a small amount of salt—most commonly sodium chloride. The salts inside your cells and the salts in your blood stay in a perfect balance called osmolarity. If there is a lot of salt in your cells, water from the blood rushes into the cells and swells them until there is the same concentration of salt in your cells and outside of them. The reverse is also true—if there is not a lot of salt in your cells, fluid will leak out of the cells and shrink them until the salt concentration is the same inside and outside of the cells.

In your eyes, the tears produced by glands in the upper eyelid (lacrimal glands) help to protect the eye and wash out dirt or germs. The tears are mixed with oil secreted from eyelid glands, so that the tears will not dry up too quickly. It is important for the tears to contain salt so that the cells inside the eye do not swell or shrink but maintain normal osmolarity. Thus, the eyeball does not dry out and is protected from harm.

What are growing pains?

Most people think growing pains are caused by growth spurts resulting in bone pain. This theory has never been proven. Growing pains can occur in half of all children and are most common between the ages of three and twelve. The pain is felt in both legs and it comes and goes during the evening and at night. Growing pains are not associated with limping or joint pain and do not result from trauma to the legs or infection. Some researchers believe that growing pains are just the normal aches and pains that occur after a child has been active during the day. Growing pains are not dangerous and will go away by themselves before adulthood. The doctor may recommend an over-the-counter pain medicine for comfort.

Why do babies have a "soft spot" on their head?

||||||||||||||||||||||||||||||||||||||||||||||||||||||||||||||||||||||||||||||||||||||||||||||||||||||

the skull is not just one round bone that protects your brain. It is actually several curved bones that are connected by seams. In a newborn baby, these seams or joints are very loose because the brain grows the fastest during the first two years of life. The places where the seams are the largest and most open are called the fontanelles, or soft spots. These usually close up by the end of the second year of life. Then the brain continues to grow in size until the teenage years—around fifteen or sixteen.

When you hold a newborn baby, you have to be careful to protect the baby's soft spots. The brain is vulnerable to injury in these places because the skull has not yet grown to cover the brain completely.

Why can't one eye move up while the other eye moves down?

eye movement is controlled by three pairs of muscles that surround the eyeball. They are attached to the eyeball on one end and to the bone of the eye socket at the other end. Because these muscles move in pairs, they only allow the eyes to move in the same direction at the same time. The muscles allow the eyes to look up/down, left/right, and any combination of these, such as upper left or lower right.

If one of the muscles (or the nerve to the muscle) is injured, the eyes cannot move together and the person ends up with double vision (diploplia). Some people have a weakness of one of the eye muscles—it makes them look cross-eyed (strabismus). They may need to do special eye-strengthening exercises or even have eye muscle surgery to allow them to see normally and to make their eyes look even.

Why do your eyes close when you sneeze?

Sneezing is a reflex, an automatic response that happens even when you don't think about it. If you walk into a dusty room, you cough—a reflex that allows the throat and windpipe to be cleared of the dust that is irritating it. When you go to the doctor's office for a checkup and he or she taps on your knee with a little red rubber hammer and your leg shoots out, that kick is a reflex. So when something irritates your nose like dust, dirt, or allergies, you respond with a sneeze reflex.

The sneeze is divided into three parts. First your eyes close. Then you take a deep breath. Finally, that breath is blown out forcefully through your nose and mouth along with any dust or dirt that started the sneeze. The brain controls this reflex in its "sneezing center," which is deep in the part of the brain called the medulla. You can keep your eyes open when you sneeze, but it's extremely hard to do because the reflex causes them to shut right before the sneeze starts.

Did you know that about a third of all people have a photic sneeze response (PSR)? People with PSR sneeze in response to bright lights—if they walk outside from a dark room into the bright sunlight, for instance. No one knows exactly why the light causes them to sneeze, but the PSR has been described in medical literature for more than fifty years.

Why do men pee standing up and women pee sitting down?

both men and women can pee either standing up or sitting down, but it's most convenient for women to sit and men to stand. Urine is collected in the bladder and passes through a small tube called the urethra to exit the body. In women, the external opening of the urethra is found inside the body, in front of the vagina. When a woman sits, the genitals open wide, allowing the urine to pass directly into the toilet. For a man, the urethra passes through the penis, which is on the outside of the body. It is easier for him to stand and aim the urine into the toilet. If his aim is not good, he should pick up the toilet seat to prevent a mess!

Why can't men become pregnant?

||||||||||||||||||||||||||||||||||||||||||||||||||||||||||||||||||||||||||||||||||||||||||||||||||||||||||||||||||||||||||

there are many reason why men cannot become pregnant, including a lack of female hormones, but the main reason is that men do not have wombs. When a woman is pregnant, the baby lives inside her womb, or in her uterus. In the womb, the baby is safe and protected. The womb also provides a connection between the baby and the mother—the baby's umbilical cord is connected to the placenta (a special pregnancy organ containing many blood vessels), which is attached to the wall of the womb. When the mother eats, vitamins and nutrients are absorbed into her bloodstream, then passed into the placenta, through the umbilical cord, and into the baby. Any waste products produced by the baby's metabolism are then passed the opposite way, through the umbilical cord, into the placenta, and into the mother's bloodstream, where she can dispose of it.

Because men are not born with a womb, they cannot house a developing baby and they cannot become pregnant.

Why do some men go bald?

|||||||||||||||||||||||||||||||||||||||||||||||||||||||||||||||||||||||||||||||||||||||

the majority of men will experience some hair loss in their life-time, but some men go bald sooner than others. This early hair loss is called male pattern baldness, or androgenetic alopecia. It can happen in women too, but much more often in men. No one knows exactly why baldness happens, but it has something to do with male hormones and genetics—meaning other family members are bald too.

Male pattern baldness starts out with hair loss in the front, by the temples, as well as at the top of the head or crown. Eventually this hair loss can progress to total baldness on the top of the head, with just a ring of hair around the sides and back. Male pattern baldness usually starts in the mid-twenties and progresses at a different rate for each person.

Although there is no known cure for baldness, there are some medications that can be applied to the scalp as a lotion or foam to slow hair loss and sometimes regrow areas of hair. Some men choose to have hair transplants—a procedure in which a surgeon takes hairs from the back and sides of the head and transfers them to the bald parts. Still other men cover their baldness with a wig or hairpiece.

What does it mean to be tongue-tied?

the dictionary definition of tongue-tied is to be too shy or embarrassed to speak, but there is a medical definition of tongue-tied as well. If you look in a mirror and lift your tongue, you will see that it is connected to the floor of your mouth by a thin slip of tissue called the frenulum. If the frenulum extends all the way to the tip of the tongue, it can tether your tongue to the floor of your mouth, preventing you from sticking out your tongue past your lower gum. This tethering is called ankyloglossia.

A little less than 10 percent of all newborns have some degree of ankyloglossia. If it is severe, the infant may be unable to breastfeed, or an older child may have speech impediments or have difficulty brushing the teeth. In these extreme cases, the doctor can perform a frenotomy, which means making a small slit in the frenulum to release the tongue to solve the problem.

Why do people get dandruff?

the medical term for dandruff is seborrheic eczema. It is caused by a fungus that lives on the skin called malassezia. In some people, this fungus or yeast irritates the skin, forming white flakes of dead skin and oily patches. Dandruff usually occurs on the scalp (in babies, it is called cradle cap), but it can also occur on the body, face, eyebrows, or ears. The symptoms include itching, flaking, and greasiness of the skin. Dandruff occurs in about 3 percent of the population.

Researchers don't understand why some people get dandruff and others don't, but they do know that if your family members have dandruff, you are more likely to get it. Other factors that affect dandruff include stress, cold weather, dirty skin or hair, and use of harsh shampoos or soaps. If you have dandruff, it's easy to treat with special shampoo that you can buy in any drugstore.

Body Afflictions and Everyday Strange Things That Can Happen

t his part covers many of the problems that can happen to your body—from mosquito bites to poison ivy. If you have food allergies, color blindness, or sunburn, you can find the answers here. In this part, we review why your ears pop on a plane and why blond hair turns green in the pool. If you've ever wondered why you can't get rid of a song stuck in your head, we'll give you the answer.

Why are some people color-blind?

many people believe that if you are color-blind you see the world in black and white like a black-and-white TV or photograph, but that type of color blindness (called achromatopsia) is extremely rare. The most common type is the inability to distinguish red-green, which are seen as white or gray. Color blindness is more common in men, and about 6 to 9 percent of white men have this type of vision problem. People who are color-blind are usually that way because they inherited the color blindness, but some diseases can cause it as well. There is no treatment and no cure for color blindness.

What makes some people stutter?

tuttering affects about 1 percent of the population in the United States. Although there are different types of stuttering, they are all characterized by speech that is not fluent. For example, instead of "I want to play baseball," a person who stutters may say, "I-I-I-I want to play baseball." They may also pause during a word—so that the word "computer" comes out as "comp . . . uter." Other examples of stuttering include adding "ums" and "uhs" to sentences, such as "I uh like dancing" or dragging out words, like "Suuuuuummmer is my favorite seeeeeeason."

The cause of stuttering is multifactorial, including genetics (if your mother or your father had problems with stuttering, you may as well), child development, brain development, and environmental influences such as rushed, stressful conditions.

Most people who stutter begin before they are ten years old, and more boys stutter than girls. Some people with stuttering can stutter a lot of the time, others stutter only a little. If people are very nervous, as when giving an oral presentation at a conference, they may stutter a lot, but when they are at home singing a song or talking to their pet, they may not stutter at all. There are many treatments for stuttering and, with speech therapy, many people may improve or stop stuttering.

Why do some people need to wear eyeglasses?

||||||||||||||||||||||||||||||||||||||||||||||||||||||||||||||||||||||||||||||||||||||||||||||||||

m ore than 100 million people in the United States wear glasses, and four main conditions cause them to need help seeing. First is nearsightedness, also called myopia. With this condition a person can see well up close—as when reading a book—but cannot see far away. The opposite problem is called far-sightedness, or hyperopia, when a person can see a movie screen clearly but cannot read a newspaper. Presbyopia, the inability to focus on near objects, occurs as we age, usually in people over forty. People with presbyopia need glasses to read a book or news-paper. Last, some people need vision correction because the front of their eyes (the corneas) are irregular. This problem is called astigmatism.

● ● ● ● ● ● ● ● ●

Why do some people have food allergies?

Your body's immune system is designed to fight infection, but sometimes it functions incorrectly. When someone has a food allergy, the body seems to think that food is bad, and the immune system reacts accordingly—usually with hives (itchy red bumps on the skin), diarrhea, tongue swelling, or closure of the airway (also known as anaphylaxis), which can be life threatening. Food allergies run in families. If your parents have food allergies or other allergic conditions such as hay fever or asthma, you are more likely to develop food allergies than people whose parents are not allergic. Some people may outgrow their food allergies, but many do not—these people need to refrain from eating the foods that cause them to react negatively.

More than one hundred foods have been identified to cause allergies, but the vast majority of food allergies come from eight food families: milk, eggs, peanuts, tree nuts (such as almonds, cashews, macadamia nuts, pine nuts, pistachios, and hickory nuts), fish, shellfish, wheat, and soy. People with food allergies need to carefully check the labels of all the foods that they eat in order to avoid a problem. You probably know someone who has a food allergy to peanut butter. Many schools have become "nut free" in order to help children with this common problem. Although there is currently no known cure for food allergy sufferers, this field is an active area of medical research.

Why, when I'm asleep at night, do I sometimes startle myself awake?

a s you pass through light and deep sleep cycles during the night, your arms and legs may occasionally twitch or jerk. Sometimes the movement is enough to wake you up briefly. If you have clusters of these repeated leg jerks—at least five jerks an hour—you may have periodic limb movement disorder (PLMD), which is characterized by leg jerks approximately every twenty to forty seconds, each of which causes a brief awakening.

Restless leg syndrome (RLS) is another type of motor syndrome that can cause sleep problems. RLS is characterized by a creepy-crawly or pins-and-needles sensation of the lower legs, which is only relieved by moving them. Both PLMD and RLS can cause difficulty falling asleep and staying asleep, which can lead to daytime sleepiness. In moderate to severe cases, prescription medication may be helpful.

Why do your ears pop when you're on a plane?

Your ear can hear because your eardrum moves when it catches air vibrations or sound. Your ear is connected to the back of your nose and throat by the auditory or eustachian tube. This little tube allows air to pass through the ear, and the eardrum to move freely. When you are on a plane, especially one that is landing, the pressure inside your ear may be less than the cabin pressure. Thus the air presses down on the eardrum, causing pain. If you chew gum or yawn, you can open the eustachian tube and help equalize the pressure between the plane and your ear. So start chomping if you don't want to have ear pain while you're flying!

Why do you get sunburned?

||||||||||||||||||||||||||||||||||||||||||||||||||||||||||||||||||||||||||||||||||||||||||||||||||||||

Y ou get sunburned by staying out in the sun for too long, especially without sunscreen lotion. The sun contains two types of light, called ultraviolet (UV) rays—UVA and UVB. UVA rays can pass deep into the skin and cause certain types of skin cancer. UVB rays stay on the surface of the skin and cause sunburns. Fair-skinned people are more likely to get sunburned than dark-skinned people because their skin has less protective pigment (called melanin). When the skin is exposed to strong sun without protection, the sun's rays can destroy the top layer of skin, causing redness, pain, and sometimes blistering.

Everyone should wear protective clothing (like a hat) when out in the sun, and those over the age of six months should wear sunscreen. Sunscreen lotion is labeled according to its SPF or sun protective factor. The higher the SPF, the more protection it provides. If you are going to be outdoors in the sunlight, you should wear an SPF of at least 15, meaning that the lotion provides you with fifteen times the protection from the sun versus wearing no lotion at all. You should put on the lotion at least fifteen to thirty minutes before going outside, so it has a chance to penetrate the skin evenly. Reapply your lotion every two hours or more frequently if you are swimming. Waterproof sunscreen lotion is available for swimmers. Avoid being outside between ten a.m. and two p.m., when the sun is the strongest.

If you do get burned, do not apply ice, as it can further destroy the skin cells by being too cold. Instead, apply over-the-counter medicine and take nonsteroidal anti-inflammatory medicine (NSAID) to ease the pain. If your burn is severe, you will need to see your doctor for prescription medicine. It's best to be careful in the sun, because severe sunburns in childhood have been linked to skin cancer later on in life.

Why do you sometimes get a stitch in your side when you run or walk fast?

everyone is familiar with the cramping pain that you some-times get during running—it is called a stitch or, in medical terms, exercise-related transient abdominal pain (ETAP). There are many theories as to why you get ETAP—some researchers think it is because during exercise you don't get enough blood flow to your diaphragm, the muscle that separates your chest from your abdomen. Others believe that irritation of the abdominal lining can cause pain. Finally, some researchers feel that the jostling movement of running can bounce the spinal bones (vertebrae) and cause pressure on the spinal nerves, leading to pain. Horseback riders are especially prone to ETAP, possibly because of all of the bumping while riding.

You may be able to decrease the incidence of ETAP by stretching before you work out, drinking plenty of water, breathing evenly and slowly during exercise, and avoiding a large meal before you head out to the track.

Why do people sleepwalk?

t he medical term for sleepwalking is somnambulism; it is one of various sleep disturbances called parasomnias. Others include bedwetting or sleep terrors (frightening nighttime episodes). Somnambulism is seen in about 15 to 20 percent of children between the ages of four and eight years. It usually goes away on its own by adolescence but sometimes persists into adulthood. It is more common in children who have had family members who were sleepwalkers too.

Somnambulism usually occurs during the first hour or two of sleep, just as one is making the transition from deep sleep to dreaming. Sleepwalkers leave their beds, their room, and sometimes even the house! They can perform complicated tasks such as locking or unlocking the door, cleaning the house, or cooking food. Episodes usually last less than ten minutes, and the sleepwalker has no memory of the events. Because most children grow out of sleepwalking, no specific treatment is recommended, other than to keep a safe environment to prevent accidental harm. For people with persistent problems, sleeping medications, psychological therapy, and waking the person just before the sleepwalking episode is about to occur usually are successful remedies.

Why does blond hair turn green in a pool?

I t's not the chlorine in the pool that makes your hair turn green—it's the tiny amounts of copper. This metal gets into the pool water from metal pipes and equipment, as well as from chemicals put into the water to prevent growth of bacteria and algae. The hair absorbs the copper, which leaves a greenish residue. This green color is most easily seen in people with blond hair, but brunettes and redheads have the same green residue. You can minimize this effect by rinsing your hair immediately after swimming (rather than allowing the pool water to dry on your hair) and by using special shampoos and conditioners designed for swimmers to try to get rid of the green buildup.

Why does your heart beat faster
when you exercise?

Your heart is a pump. Veins leading into the heart deliver blood that is low in oxygen to the heart. The heart then pumps this blood to the lungs, where it picks up oxygen. This oxygenated blood is delivered back to the heart, and the heart pumps it out to the body via a system of tubes called arteries. The body needs this oxygen to function—for example, to keep your muscles working and your brain thinking. The amount of blood delivered to the body is a combination of how much blood volume is in each heartbeat multiplied by how fast the heart is beating. In other words, if the body needs more oxygen quickly, the heart has two choices—to pump more blood in each heartbeat or to pump it faster. Since the heart is a muscle, it can enlarge a little bit to accommodate more blood in each beat, but it can really get more oxygen to the body by pumping faster.

Put your hand over your heart, and you can feel it beating. You can count the number of beats in a minute. Now do some jumping jacks for five minutes. When you stop, count the number of beats per minute again. You can really feel your heart pounding after exercise because it is pumping out more blood with each beat and doing it at a faster rate to get the oxygen to your body's muscles.

Aerobic exercise is essential to make your heart beat faster and stronger. It helps to keep the heart healthy.

What is athlete's foot?

Y ou don't have to be an athlete to get athlete's foot—you just need feet. Athlete's foot, or tinea pedis, is a foot infection caused by a fungus. The fungus likes to live in a warm, moist environment, like a locker-room floor; hence the name athlete's foot. You can get athlete's foot by having your foot touch the fungus—on the floor, in your shoes, or by touching your foot on someone else's foot. Athlete's foot is an itchy, cracked rash that occurs on the foot, especially between the toes.

It is easy to catch athlete's foot, but it is also easy to get rid of it. Antifungal medication is sold at the drugstore without a prescription and can be applied by cream, lotion, or spray. Athlete's foot can return if you have contact with the fungus again, so always wear shoes or flip-flops at the pool or in public showers. If your shoes get wet, allow them to dry thoroughly before putting them on—the fungus thrives in smelly tennis sneakers wet with sweat.

Why do your lips turn blue
when you swim in cold water?

Cold-blooded animals, like lizards, have a core temperature that is dependent on the outside temperature—if a lizard is sitting on a warm rock, the lizard's body temperature is warm. If the lizard is in the cool shade, its temperature is cool. Human beings are warm-blooded animals. Our temperature stays at around 98.6 degrees Fahrenheit whether it is cold or hot outside.

Human beings have many ways to maintain their temperature at its set range. One way we do this is by blood flow. If you are swimming in a cold pool or lake, the blood vessels near the surface of your body (lips, hands, feet) constrict and get smaller. This reaction allows more of your blood, which is warm, to flow to your inside organs and keep them warm. Because less blood goes near the surface of the body, less heat escapes through the skin. Shivering also helps keep you warm by creating energy, which warms up the body. When the red blood flows away from the surface, your extremities, like your lips and the tip of your nose, feel cold and appear blue. The scientific name for this is cyanosis. When you finish swimming and sit in the sun, the blood vessels near the surface expand, filling them with warm red blood, and your normal complexion comes right back.

Why do mosquito bites itch?

osquitoes are found in every part of the world except Antarctica. Only the female mosquitoes bite because they need blood in order to lay their eggs. As the mosquito bites you, small amounts of mosquito saliva (otherwise known as mosquito spit) enter your skin and cause the body to produce an immune reaction to fight off the irritating chemicals found in the mosquito saliva. This immune reaction causes a small red bump to form at the bite site and is accompanied by itching and mild pain. You can use over-the-counter creams or lotions to help soothe a mosquito bite, but try not to scratch because then it can become infected. In addition to being annoying and itchy, mosquitoes can spread diseases, so it's a good idea to wear mosquito repellent when you will be outdoors in places where mosquitoes like to live—near water and swamps.

Why do some people grind their teeth at night?

nighttime tooth grinding is called bruxism. It is characterized by tooth grinding or jaw clenching during sleep. Signs and symptoms of bruxism include abnormal tooth wear, tooth fractures, jaw pain/tenderness, jaw muscle discomfort, and headaches. The tooth grinding disrupts a person's sleep—as well as their sleep partner's sleep—due to the noise associated with the teeth rubbing together. Around 15 percent of children and about 8 percent of adults grind their teeth. Risk factors for bruxism include anxiety or stress, as well as smoking or alcohol use. People who have family members affected by bruxism have a higher likelihood of bruxism themselves. Treatment for bruxism includes oral splints to separate the upper and lower teeth, sleeping medications, and behavioral therapies to reduce stress and promote relaxation.

Why does your face pucker when you suck on a lemon?

Your face puckers up because the lemon contains citric acid which stimulates the production and flow of saliva—spit. You need saliva to lubricate and break down the food that you chew so it can glide into your stomach easily. The saliva is made in the salivary glands—these are little glands under your jaw and tongue, but the largest is the parotid gland, which is located in front of your earlobes on both the right and left sides of your face.

Once the saliva is made by the parotid gland, it travels into your mouth via a tiny tube called Stenson's duct. The tube opens up into the mouth through a small opening. If you shine a flashlight into someone's mouth and look at the inside of the cheek—opposite the top row of teeth in the back, you can see a little bump, which is the opening of Stenson's duct. Once you find the duct, ask the person to suck on a slice of lemon—you can actually see the saliva come out of the duct. The person's face will start to pucker as the muscles around the parotid gland squeeze the saliva through the tiny tube into the mouth.

Patients sometimes have dry mouths from medication or illness. Without saliva, it is difficult for them to chew and swallow their food. The doctor may prescribe lemon drops to improve the flow of saliva and relieve their symptoms.

Can your contact lens get lost in your eye and float into your brain?

no! Your eyelids, both top and bottom, are connected to the eyeball itself, thus preventing the contact lens from drifting behind your eye. If your contact lens is not centered over the front of your eye (the colored part, or the iris, and the dark circle, or the pupil), it cannot help you see. Sometimes the contact can fold up on itself and drift to the side of the eye. By gently feeling the surface of your eye, you can usually find out where it is. Sometimes someone else may have to look at your eye while you are pulling back on your lids in order to find the missing contact.

Why do some people get fat?

||||||||||||||||||||||||||||||||||||||||||||||||||||||||||||||||||||||||||||||||||||||||||||||||||||||

Obesity is an increasing problem in this country. More than 60 percent of adults and an alarming number of children are overweight or obese in the United States. Besides being a social stigma, obesity has many serious medical consequences, including high blood pressure, diabetes, and high cholesterol.

The main reason that people are obese is because they eat too many calories and don't get enough exercise. Unhealthful high-fat and high-calorie foods at fast food restaurants are becoming increasingly inexpensive, while healthful foods, such as fresh fruits and vegetables, are oftentimes more expensive. Although genetics has a role (if your parents are obese, you are more likely to be obese), environment also contributes. If your parents serve you unhealthy foods in large quantities, you are more likely to continue these bad habits into adulthood.

Getting exercise is important in preventing obesity. Many adults have jobs that involve very little physical activity—sitting at a desk or in an office all day long. Kids watch a lot of TV or play computer games rather than exercising. This sedentary lifestyle contributes to obesity.

Some people are overweight because of medical problems like an underactive thyroid. But most people are obese because they eat too much and exercise too little. It is hard to make lifestyle changes, but it is essential to your health to maintain a normal weight for your height and body type.

Why do mosquitoes bite some people and not others?

||||||||||||||||||||||||||||||||||||||||||||||||||||||||||||||||||||||||||||||||||||||||||

female mosquitoes are the ones that bite and need human blood to survive. The male mosquitoes eat flower nectar. The females typically feed every three to four days and eat their own weight in blood!

Mosquitoes use at least three factors to find someone (or something) to bite. The most important factor is smell. Your body is constantly burning energy and, as a result, making waste products like carbon dioxide and lactic acid. Mosquitoes can sense these waste products and use their sense of smell to find you. Sweat can also attract them. If you use shampoos, soaps, or laundry detergents with heavy fragrances, these odors can sometimes cover up your body's natural scent and thus repel the mosquitoes. But sometimes these fragrances can have the opposite effect and actually attract the mosquitoes. These insects also use sight and temperature to find prey. Movement and large size attract the mosquitoes and warmth can also draw them in. If you wear dark colors, which trap the heat, it's like giving the mosquitoes an open invitation. Wear light-colored long sleeved shirts and pants and use insect repellents to avoid getting bitten.

What is a hangnail?

Your nail begins at the base and grows up to your fingertip. Skin surrounds the outside of the nail, forming a U-shaped seal around it. This skin is commonly referred to as your cuticle. If this skin becomes ragged or torn, it forms a hangnail. The hangnail then opens up a small space in the seal around your nail. If you bite your nails or suck your fingers, a bacterial infection can be introduced around the nail through this break in the seal. This type of infection, a paronychia, may need to be treated with antibiotics, or your doctor may have to make a small incision around the cuticle to drain out any pus.

People who work with their hands, such as bakers or dishwashers, are at increased risk for infection because the constant handwashing causes the cuticles to dry out and become prone to hangnails, which in turn can lead to infection. If you develop a hangnail, you should use a clean manicure tool to cut it off at the base. Do not bite or tear it off with your teeth, as your mouth can contaminate the skin and cause infection.

Why can't I get rid of this song that is stuck in my head?

I t happens to everyone: all of a sudden, for no reason at all, you start singing a song in your head. You may not even like the song, and you may find it annoying. But there it is, stuck in your head for a couple of minutes, hours, or even days. The song is inside your head—not a voice singing it, but the words unavoidable in your mind.

There are many names for this type of phenomenon, including last song syndrome, earworm, or repetunitis. There are no specific requirements for the song—although it is usually simple and the lyrics are repetitive. Researchers do not understand why last song syndrome occurs. We know that your thoughts can be formed through hearing (phonological loop). Perhaps something malfunctions in this loop, allowing the song to be replayed in your head. Some researchers liken it to a brain itch that needs to be scratched, but the more you scratch, the itchier you get—or the more the song keeps on playing in your mind.

There's no definite way to get rid of last song syndrome. Sometimes if you just sing the song out loud from beginning to end, it can stop the repetition. If you focus on how annoying the song is, it usually keeps replaying. Try to concentrate on something completely different. If you don't actively try to resist the song, it may just go away as you deal with a new topic.

Why does poison ivy make us itch?

Poison ivy is a plant. Its leaves grow in clusters of threes. Poison ivy produces an oil called urushiol (also found in poison oak or poison sumac), which is irritating to your skin. The type of itchy, raised red rash that you get is called allergic contact dermatitis. If you come into contact with poison ivy, wash your skin as soon as possible to try to get rid of as much oil as you can. If your clothes have touched the poison ivy, do not touch your clothes, because you can transfer the oil from your clothes to your skin. Once you get the rash, you are not contagious to other people because the oil has already been absorbed into your skin. Although you can use creams and lotions to help stop the itching, the best way to prevent a poison ivy rash is to avoid the poison ivy plant. So remember, "Leaves of three, do *not* touch me!"

Urban Myth and What If?

has your mother ever told you to stop cracking your knuckles or you'll damage your joints? Or how about, "Don't sit too close to the TV or you'll hurt your eyes"? Is it really necessary to wait an hour after eating before going swimming? Well, parents don't always have the correct answers. In this part, we investigate some of the most common adages to determine whether they are fact or fiction. We also tackle some urban myths, including the one about your stomach exploding if you mix Pop Rocks and soda.

Is it really bad to crack your knuckles?

We are not certain what causes the popping sound when you crack your knuckles. The knuckles or finger joints are two bones that are held in place by ligaments and tendons and lubricated with fluid (synovial fluid), much like the pistons in your car engine are greased with oil. When you bend or move your fingers in unusual ways, like bending them backward instead of forward, you can force the fluid to move rapidly in the joint space, creating bubbles that may produce the popping sound. No one has ever linked cracking your knuckles with arthritis or any other joint ailments, but, like nail biting, it is an annoying habit that you should try to break.

What happens if you swallow a penny?

The mouth is connected to the esophagus—a long tube—which leads to the stomach and then the intestines and out the rectum. So, if the foreign object that is swallowed is small enough, it will pass out in your stool in a day or two. But if the object is too large, it will get stuck. Sometimes it will get lodged in the esophagus or stomach, making it painful or difficult to swallow food. Then you need medicine to put you to sleep (anesthesia) and the doctor takes a long, lighted telescope (an endoscope) and pulls out the penny or whatever was swallowed. Andrea and Julia's Uncle Michael is an ear, nose, and throat doctor, and he says that if he could keep all of the coins that he has taken out of kids' esophagi through the years, he would be very rich! In some cases, the doctors have to take an X-ray every day so they can see if the penny is moving along through the intestines and not getting stuck anywhere. Then you may have to look through each one of your bowel movements to see when the penny finally comes out. Yuck!

What happens if you eat dog food?

IIIIIIIIIIIIIIIIIIIIIIIIIIIIIIIIIIIIIIIIIIIIIIIIIIIIIIIIIIIIIIIIIIIIIIIIIIIIIIIIIIIIIIIIIIIIIIII

I f you just eat a little bit of dog food, probably nothing will happen. But you don't really know that for sure because dog food is not subjected to the same health and safety regulations that human food is required to have. In the United States, the Food and Drug Administration (FDA) is responsible for setting human food safety guidelines to prevent sickness due to contaminated or improperly handled products. In addition, the FDA is responsible for ensuring that food product labels are accurate in terms of ingredient listing and nutritional content.

Dog food does not have these same strict rules that human food has. Although most dog food contains the same basic components that are in people food—protein, carbohydrates, and fats—the proportions of these ingredients are different than in human food and can be harmful if ingested in significant quantities or for prolonged periods of time. The same is true if you feed a dog the wrong proportion of these nutrients by giving it an unbalanced human diet. Dog food contains many of the same ingredients as human food, like chicken, meat, and vegetables, but it may also contain animal by-products—for example, ground-up animal bones or organs like the intestines. The best advice is to keep dog food for dogs and human food for humans!

Can you tell your fortune by reading your palm?

||||||||||||||||||||||||||||||||||||||||||||||||||||||||||||||||||||||||||||||||||||||||

or centuries, fortune-tellers have tried to look into the future by examining the creases on your palms, sometimes claiming that the lines will tell them how long a life you will have. Unfortunately, there is no scientific evidence that this is possible. The three longest lines on your palm are the radial longitudinal crease (the C-shaped line that arcs around the base of your thumb) and as the two horizontal creases; the distal transverse crease (the higher line); and the proximal transverse crease (the lower one closer to your wrist).

Dermatoglyphics is the science of looking at these palm creases. We all have these natural skin folds that allow the skin to bend and stretch as we move our hands. Some genetic diseases produce a characteristic change in the normal pattern of creases—for example, in Down's syndrome there is only one horizontal crease instead of two. Doctors use these abnormal patterns to help them diagnose some diseases that are passed down in families.

Why do you get a headache if you eat ice cream too quickly?

||||||||||||||||||||||||||||||||||||||||||||||||||||||||||||||||||||||||||||||||||||||||||||||||||

S cientists have actually researched this question of "ice cream headaches," otherwise known as brain freeze. About a third of people (especially children) will get a headache when they eat ice cream too quickly. An ice cream headache is different from tooth sensitivity, which is caused by receding gums, allowing the root of the tooth to be exposed to the cold ice cream and triggering tooth pain. Instead, an ice cream headache is usually in the forehead and temples, and it lasts for less than a minute. Scientists think that the cold ice cream causes decreased blood flow to the brain, thus prompting a headache—perhaps similar to the physiology of a migraine headache. Although brain freeze is annoying, it doesn't stop anyone from finishing their ice cream. One folk remedy is to press your thumb to the roof of your mouth—maybe this maneuver warms up your mouth or possibly just distracts you temporarily from your headache.

Can you really get rabies from a dog bite?
What is rabies, anyway?

abies is a viral type of infection seen in mammals, most frequently wild animals like skunks, bats, and foxes. Animals with rabies appear sick—sometimes drooling or snapping or acting aggressively. Of the millions of people who are bitten by animals every year, most of these bites are from dogs. Pets can get rabies from wild animals and then can transfer the rabies to humans by biting them. The veterinarian can give your pet a rabies shot to prevent this infection.

People can get rabies from bats by being in close contact with them and breathing in the rabies virus. Anyone who has been bitten or comes into close contact with a rabid animal should seek medical attention immediately to get special shots to prevent them from catching the rabies virus. The shots hurt a lot, but people who contract rabies will die, so it is important to be as careful as possible.

What happens if you swallow gum?

||||||||||||||||||||||||||||||||||||||||||||||||||||||||||||||||||||||||||||||||||||||||||||||||||||||||

most gum is chewed by children, and although it can cause cavities and gum disease and sore jaws, most kids just keep right on chewing. Adults love gum too, and although most people throw away their gum in the trash when they are finished chewing, some people swallow it on purpose or by accident.

Most of the time, nothing bad will happen if you swallow a small piece of gum. It will just pass through your intestines and get mixed in with your bowel movement and leave the body that way, usually in about twenty-four hours. But if you chew and swallow a lot of gum, the gum inside your throat or in your stomach can form a big, sticky wad. If this lump of chewed-up gum becomes big enough, it can actually block your throat, making it difficult to swallow, or it can obstruct your intestines, making it impossible for a bowel movement to pass through your rectum. If either of these situations happens, you may need surgery to pull out the wad of chewing gum. If that's not enough to make you stop swallowing your chewing gum, then I don't know what is!

If you put a pea up your nose will it go into your brain?

n o. Your nose is connected to your brain through nerves that travel from the smelling cells in the nose to the olfactory bulb—the part of the brain that allows you to smell. A pea can't travel through nerves. If you put a pea up your nose, it usually lodges in the floor of the nose, which is narrow. If the pea is very small, it can travel to the back of the nose, and down your windpipe into your lungs. If this happens, a doctor may need to remove it using a small lighted instrument that looks like a thin telescope.

Little kids often put things up their nose, even though they know they shouldn't. Most common are rubber erasers, balls of paper, pebbles, marbles, beans, and peas. Sometimes living things, like maggots or worms, can get wedged in the nose, which is really disgusting. If a child is very little, he may not be able to tell his mother what he did. Instead, he may have a nosebleed, or pus and a foul smell coming from his nostril. Most things that get stuck up the nose can be removed in the doctor's office with forceps, clamp, or hook. Only rarely does a child need to have surgery to remove it.

Can you catch a cold if you don't wear a coat in winter?

||||||||||||||||||||||||||||||||||||||||||||||||||||||||||||||||||||||||||||||||||||

You may be surprised to learn that there is no scientific evidence that going outside in a snowstorm without your winter coat will cause a cold, but it's still not a smart thing to do. The same is true for going outside with a wet head—your hair might freeze in the cold and become damaged or even break off.

Although there is some evidence that exposure to cold temperatures may decrease the function of your immune system, common colds are caused by viruses, not by a decreased immune system. There are more than two hundred different kinds of viruses linked to the common cold. There are about a billion cases of colds in the United States every year. The cold viruses infect the lining of your nose. Then your body's immune system responds to this infection by creating the cold symptoms we are all so familiar with—sneezing, coughing, congestion, and an itchy, runny nose. Wearing a jacket or a hat in the winter will not prevent you from getting a cold, but it will make you feel warmer. There are three important ways to prevent a cold. First, wash your hands, especially when you come into contact with someone who has a cold. Second, avoid touching your eyes and nose—this will prevent any viruses on your hands from spreading to your nose. Last, avoid other people when they are coughing or sneezing. This may be hard to do in a classroom or tight space, like inside an airplane. If you cover your nose and mouth when you cough and sneeze, you can prevent spreading your cold to other people.

Can your MP3 player really cause you to go deaf?

||||||||||||||||||||||||||||||||||||||||||||||||||||||||||||||||||||||||||||||||||||||||||||||||||||||||||||||||||||||

listening to loud music for long periods of time can cause irreversible hearing loss secondary to damaging the delicate hair cells in your inner ear (cochlea). Since teenagers are most likely to listen to loud music, they are most at risk. People who prefer ear-bud-style headphones rather than the over-the-ear variety hear more background noise and may increase the volume on their MP3 player in order to hear the music better.

Signs of hearing loss include: (1) Thinking that people are mumbling or not talking clearly when in fact they are speaking normally; (2) increasing the volume on your TV or radio when others can hear it clearly; (3) frequently asking people speaking to you to repeat themselves because you did not hear them clearly the first time; and (4) ringing in your ears. Although it may be fun to listen to loud music in the short term, even small exposures to very loud music can cause permanent damage. So be smart and turn down the volume!

What would happen if I never brushed my hair?

||||||||||||||||||||||||||||||||||||||||||||||||||||||||||||||||||||||||||||||||||||||||||||||||||||||||

f you never brushed your hair, it would get very, very knotty. Eventually, these knots would become matted together and form long ropes of hair (sometimes referred to as dreadlocks). This process is similar to felting wool. If you put a wool sweater into the washing machine and agitate it for several minutes, the wool will turn into felt.

Many people wear their hair in dreadlocks for religious, ethnic, or cultural reasons. The coarser and curlier your hair, the faster the knots will turn into dreadlocks. For people with very straight and fine hair, dreadlocks may not form for several years of not brushing. Some people think that by not washing their hair, it will form dreadlocks sooner—but this myth is not true. If you don't wash your hair it will get dirty, smelly, and unhygienic. Dreadlocks can be washed just like washing a sponge. After soaping, repeated rinses and gentle squeezing will eliminate the soap and residue.

If you want dreadlocks more quickly, you can divide your hair into sections and then tease (or back-comb) it in order to form the matted ropes. There are also commercially available dreadlock perming kits to make your hair kinkier and allow the dreadlocks to form in a shorter period of time.

If you drop food on the floor, is it still safe to eat if it doesn't touch the floor for more than five seconds?

the answer is no. Once the food touches the floor it can be instantly contaminated with illness-causing bacteria. Although many floors may be quite clean, you never know, since you walk on the floor with shoes worn outside. Pets can also spread germs onto the floor. Germs can be found on all types of floor surfaces, including wood, tile, and carpets. If the food or the floor is wet or moist, bacteria is more likely to be transferred.

Researchers have studied this "five-second rule" and have found that people are unwilling to throw away dropped food if it is sweet—candy and cookies are more likely to be picked up and eaten, while broccoli and cauliflower are more likely to be picked up and thrown away. Remember also that the floor is not the only dirty kitchen surface—countertops and cutting boards can be contaminated, especially if they've been cleaned with a dirty sponge or wiped with a germy dish towel. Use common sense and good hygiene (wash your hands with soap and water) when preparing and handling your food.

Is it true that a dog's mouth is cleaner than a human's?

many people think their dogs are so clean that they let them lick their faces. Before we even go to the scientific answer for how clean a dog's mouth is, let's just use a little common sense. Have you watched a dog recently? Our dog, Bob Dylan, loves to smell and lick things that he finds—in the yard, in the trash, other dogs, or even his own private parts. He picks up germs from all that sniffing and licking, and I certainly wouldn't want him to spread them all over my face—particularly my lips!

People think that dog mouths are cleaner than human mouths because dogs often lick their wounds. So people assume that dog saliva is clean and possibly even has healing properties. That is just not the case. Dog mouths have as much bacteria as human mouths do, but they have different types of germs than humans have. Although some of these germs are specific to dogs only and can't be passed to humans, other germs can spread from dog to human. The reason why a dog licks its wounds is because the dog's rough tongue clears away any dirt or dead tissue in the wound, making it heal faster.

Since there are hundreds of thousands of emergency room visits for dog bites in the United States every year, you should be careful of allowing a dog to come very close to your face. Almost half of all dog bites happen to children, particularly young children less than five years old. Dog bites can cause infection, pain, and scarring. My advice is to cuddle up with your dog and give him lots of love, but don't let him kiss you on the mouth.

If you eat Pop Rocks candy and drink soda at the same time, will your stomach explode?

|||||||||||||||||||||||||||||||||||||||||||||||||||||||||||||||||||||||||||||||||||||||||||||||||||

n o. Both Pop Rocks candy and soda are carbonated—meaning they contain small quantities of carbon dioxide, which is a colorless gas. For Pop Rocks, the candy coating forms a shell around tiny pockets of the gas. As the candy dissolves in your mouth, the carbon dioxide escapes, making a fizzy feeling on your tongue. For soda, the carbon dioxide is dissolved in the liquid. When you open a bottle of soda, the carbon dioxide is released as little bubbles that tickle your nose and mouth as you drink. When soda becomes "flat," it has lost all of its bubbles or carbon dioxide. When you eat Pop Rocks candy and drink soda together, carbon dioxide escapes and you can feel the fizzies as you swallow, but it's not enough to make your stomach or windpipe explode, and you can't die from it.

If you eat carrots, will you see better?

|||||||||||||||||||||||||||||||||||||||||||||||||||||||||||||||||||||||||||||||||||||||||||||||||||||||||||||||||

Carrots contain vitamin A. This vitamin is important in helping maintain healthy vision. If you do not have enough vitamin A in your body, you can develop decreased vision or even blindness in extreme cases. As you age, the back of your eye, called the macula, can break down and cause vision problems. If you consume normal amounts of vitamin A, you can decrease this breakdown and also help to prevent cataracts, which is clouding of the clear lens of your eyes, eventually leading to blindness.

It is possible to consume too much vitamin A, which can lead to poisoning, because the vitamin is stored in your body's fat and is not easily excreted from the body. Usually this happens if someone takes too many vitamin A supplement pills. It is hard to poison yourself with the vitamin by eating foods that have it, like carrots. But it is important to eat a balanced diet with a variety of fruits and vegetables to get all of your vitamins and minerals, not just vitamin A.

Is eating raw cookie dough really bad for you?

||||||||||||||||||||||||||||||||||||||||||||||||||||||||||||||||||||||||||||||||||||||||||||||||||||||||||||||

a lthough raw cookie dough is delicious, it is made with raw eggs, and you can get food poisoning from eating raw eggs. They may contain a bacteria called salmonella, which can give you diarrhea, vomiting, belly pain, and fever. Undercooked eggs, as in soft scrambled or sunny-side-up, may also contain salmonella. Raw meat and poultry can cause food poisoning too.

That is why it is so important to cook foods thoroughly to kill the salmonella and to use safe food handling tips. If you touch raw meat, poultry, or eggs, wash your hands well with soap and water to get rid of the bacteria. If the raw food touches the countertop, make sure to clean it before putting any additional food in that area. A cooking thermometer is recommended to check that the center of your food is thoroughly cooked, thus reducing the possibility of contracting food poisoning.

Is it true that if you swallow a watermelon seed, you will grow a watermelon in your stomach?

n o! The watermelon seed will likely be carried through your digestive tract, from your stomach to your rectum, without difficulty. It eventually will pass out in your bowel movement. Rarely the seed can get stuck in your digestive tract, specifically at the opening to your appendix. If this passageway is blocked, the appendix can swell and become infected—which is called appendicitis. If you swallow lots and lots of watermelon seeds, they can get stuck together and form a ball of seeds. This ball can then get stuck in your rectum, blocking your bowel movement from coming out. If this condition happens, the doctor has to remove the seed ball so that your bowels can work properly again. So if you swallow an occasional watermelon seed, probably nothing will happen to you. But if you purposefully eat many, many seeds, you are looking for trouble!

Do twins have the same fingerprints?

no, all human beings have unique fingerprints that identify them. Fingerprints are the tiny ridges found on the pads of your fingers. Each of these ridges is extremely small (about 0.3 millimeter wide). The ridges are arranged in three basic patterns—arches, loops, and whorls. The millions of different combinations of these patterns, along with the total ridge count, produces individual fingerprints for every human being that are too complex to be duplicated by chance—similar to snowflakes, no two of which are ever the same.

When you touch an object, the natural moisture from sweat and oils produced by your skin is left behind. The study of fingerprint identification is called dactyloscopy. This science was first used in the late 1880s in England to help identify and catch criminals. Today the science is more sophisticated and uses computer technology, but it is still based on the fact that no two people have the same fingerprints. Your fingerprints stay the same from childhood through old age, and even after death.

Can you get lead poisoning from a pencil?

most pencils don't contain lead. They contain graphite, a nontoxic substance. Therefore you can't get lead poisoning from getting stuck with a pencil. If the tip of the pencil gets lodged under your skin, however, you may develop a "traumatic tattoo"—a permanent dark dot underneath your skin.

Can you be allergic to jewelry?

Yes, you can be allergic to the metals that the jewelry is made from. The most common metal allergy is nickel. Nickel is a type of metal that is found in jewelry and in many everyday items, like coins, watchbands, pens, zippers, and keys. Women are more likely to have a nickel allergy than men, probably because women wear more jewelry than men and therefore are exposed to nickel more often.

The allergic reaction that you can get from nickel is called contact dermatitis—it is a red, raised, itchy rash. It can happen after your first exposure to the nickel jewelry or after many years. People who have piercings, like pierced ears, are more likely to be allergic to nickel, and people who have more than one piercing—like ears, belly button, mouth, nose, or eyebrow—are probably at greater risk for a nickel allergy. Once you are allergic to nickel, you are allergic for life. Although there are treatments for contact dermatitis, like creams or pills to decrease the inflammation and rash, the best prevention is to avoid jewelry that contains nickel. Look for pure gold, sterling silver, or plastic jewelry. Always keep your jewelry clean, and if you develop a rash, remove the jewelry to help the contact dermatitis go away faster.

Are green bananas bad for you?

n o—all bananas are good for you, and green bananas can be especially helpful if you have diarrhea. Bananas are grown from plants in warm climates and are available in the United States year-round. There are hundreds of varieties of bananas, and the average American eats more than twenty pounds of bananas per year!

Bananas have wonderful health benefits. They contain high levels of potassium, which is helpful to your heart, and high levels of fiber, which is good for your intestines. Green bananas also contain a starch that is converted in the gut into short-chain fatty acids, or SCFA. This substance is taken up by the lining of your intestines and improves the ability of the gut to absorb food and nutrients. Scientific studies of feeding green bananas to people with diarrhea show that the green bananas nourish the intestines and help to stop the diarrhea.

Bananas are picked when they are dark green and then eventually ripen by turning light green, then yellow, then brown-spotted, and finally black. As bananas ripen, the starch in them turns to sugar, which makes them taste sweet and creamy. Most people enjoy yellow bananas that have just a little bit of green near the stems or just a little bit of dark spotting on their skins because they are sweeter than the dark green bananas.

Do you really need to drink eight 8-ounce glasses (8 x 8) of water a day to be healthy?

||||||||||||||||||||||||||||||||||||||||||||||||||||||||||||||||||||||||||||||||||||||||||||||||||||

researchers have taken a new look at this question by reviewing old data and scientific studies to see if there is any medical evidence to back up the 8 x 8 recommendation—they couldn't find any support. No one knows exactly where this 8 x 8 recommendation began or why we have included it in current health recommendations. For the average individual—someone who is healthy, living in a temperate climate, and leading a relatively sedentary lifestyle—it has been proven that eight 8-ounce glasses of water is more than the physiological requirements.

Your body has a tight hormonal system to prevent dehydration. If the volume of water in your body falls, the hormone vasopressin helps the body to hold on to its water, rather than lose it through urination. The brain is also stimulated to produce the sensation of thirst, which allows you to add new water to your stores. In the past, caffeinated drinks like soda, coffee, and tea were thought to cause dehydration. However, for people who are used to consuming moderate amounts of caffeinated beverages, these fluids can add to the body's water level just like drinking water. Finally, for people who consume excessive amounts of water, water intoxication is a risk. If the kidneys cannot produce enough urine to expel the large quantities of water, the body's delicate salt balance can be unhinged.

The current recommendation is to drink enough fluid so that you don't feel thirsty. Drinking beyond this point is not helpful and may, in fact, be harmful.

Can you get frostbite in just a few minutes if it's really cold outside?

When you are outside in very cold temperatures, particularly under windy conditions, the body parts that are exposed can freeze or become frostbitten. Depending on the weather, frostbite can happen in a few minutes to several hours. Common areas that can be at risk are the tip of your nose, ears, cheeks, or fingers. You may not feel the frostbite happening because it is so cold outside that your skin feels numb.

The first sign of frostbite is when the skin becomes very pale and looks like wax. Then the skin itself can freeze and form blisters. This process is called superficial frostbite—sometimes referred to as frostnip. If you don't get inside and warm yourself, the deeper tissues can freeze, like the bones and muscles. In this kind of deep frostbite, the finger or tip of the nose can turn black. Once this happens, there is no reversing the frostbite and the little piece of finger or nose can die and eventually fall off. This kind of severe frostbite does not happen in most circumstances, only in extreme conditions. Outdoor athletes such as skiers and mountain climbers are at risk of developing deep frostbite.

Always be careful in freezing weather—try to cover exposed skin with a scarf, hat, or face mask. Wear gloves and thick socks. Limit time spent outdoors in extreme conditions. If you have to go outside, make sure you have a friend or family member come with you to be able to check you and help you avoid hypothermia (lowered body temperature due to cold weather) as well as frostbite.

Will you go blind if you sit too close to the TV?

no—TV will not hurt your eyes permanently. Some people prefer to sit close to the TV because they are nearsighted, meaning they see things better that are near to them, rather than far away. But watching too much TV means that you are probably not getting enough exercise, and that is bad for your body in general. Also, watching violent TV shows has been linked to attention deficit disorder in children. People who watch frightening TV shows or movies can develop nightmares and have trouble falling asleep. If you watch too much TV, particularly while sitting very close to the set, you may develop temporary eye strain or fatigue, which is cured by simply turning off the television!

Why is breakfast the most important meal?

Children's breakfast eating has been studied since the mid-1960s with the federal government's initiation of the School Breakfast Program. Research has shown that if children eat a healthful breakfast they are less likely to be absent from school. They have an improvement in their schoolwork and are more attentive and energetic than those students who skip breakfast. When children eat breakfast, they are less likely to have disciplinary problems at school and have fewer trips to the principal's office for bad behavior and decreased visits to the school nurse with complaints of stomachaches and headaches. Adults who eat breakfast are more productive at their jobs and less likely to make desperate and unhealthy food choices during the day—like candy from the vending machine. When you eat breakfast, you are less likely to get that

mid-morning hungry feeling. Andrea and Julia's favorite breakfast is pancakes and raspberries with a glass of orange juice.
What's yours?

Does a tongue piercing hurt?

||||||||||||||||||||||||||||||||||||||||||||||||||||||||||||||||||||||||||||||||||||||||||||||||||||||||||||||

Yes—it hurts to have your tongue pierced. Your tongue
is a muscle that is used to shape the sounds produced by
your throat, and if you stick a needle through it, your tongue
hurts a lot!

In the United States, the most common piercing is through the
soft part of the earlobes, but more recently, body piercing has become a form of body art. You can have just about anywhere on your
body pierced, but the more popular sites are high up on the ear
(through the hard part or cartilage), nostril, eyebrow, tongue, lip,
and belly button.

Many people have body piercings, and many problems can result. First is infection. Anytime you puncture the skin, you can
introduce germs and get an infection. Some infections are small,
but some can be big, leading to redness, swelling, and pus. If the
needle used to do the piercing is not clean, you can get someone

else's germs and diseases, like hepatitis (a liver disease) or AIDS. You can also get bad scars from piercings. Suppose you pierce your eyebrow and then a year later decide you don't want to have your eyebrow pierced anymore. The scar will still be visible in most people, even after the holes have closed up. Some people form extra-large and thick scars called keloids, which can look especially obvious.

Tongue jewelry can cause specific problems. The tongue jewelry can fall off and you can accidentally swallow it, or the jewelry can bang into your teeth and chip them. If your tongue jewelry gets caught on your fork while you are eating, you can accidently rip it out and cause bleeding. If you decide you want to have a piercing, make sure you have it done by someone with experience who uses clean practices. Make sure you understand the problems that can happen. Also, think about the future—although an eyebrow piercing may seem cool now, will it still be cool when you are forty or sixty years old? You need to give yourself time to go over all of the pluses and all of the minuses before you do anything.

If you cross your eyes will they get stuck?

|||||||||||||||||||||||||||||||||||||||||||||||||||||||||||||||||||||||||||||||||||||||||||||||||

n o. Some people can cross their eyes easily. Other people have to practice. Focus on the tip of your nose or on a pencil point just in front of your nose. Eventually you will get the hang of it. Although crossing your eyes can be annoying to others, you can't do permanent harm to yourself. Another irritating trick is to flip your upper eyelid inside out. Pull gently on your eyelashes until the lid is away from your eye, then you will be able to flip it. Because the upper eyelid is made of stiff material, it will usually stay flipped until you unflip it. Even though this trick can get you a lot of attention, I wouldn't recommend it because you can scratch your eye or dry it out.

Is caffeine bad for you?

||||||||||||||||||||||||||||||||||||||||||||||||||||||||||||||||||||||||||||||||||||||||

Caffeine is a substance found in plants. When added to food or drinks, caffeine acts as a stimulant—waking you up and providing you with a boost of energy. But there are many negative consequences to consuming caffeine. First, caffeine is addictive—meaning once you have some you want more. Your body actually needs more caffeine after a while to get that same energy boost, so you end up drinking more soft drinks or more coffee in order to get that wake-up feeling. Second, when you are drinking so much soda, coffee, or tea, you are probably not drinking fluids that are good for you, like milk, to build healthy bones. All of that soda and coffee (with cream) are loaded with calories, which can make you overweight. Because caffeine also acts as an appetite suppressant, you get less hungry for important calories, like fruits, vegetables, and whole grains.

Even though we know that too much caffeine is bad for you, surveys show that the majority of children in this country have caffeine at least once a week, and most of this caffeine is coming from soft drinks. Too much caffeine can make you nervous, cause trouble falling asleep, and increase your blood pressure and your heart rate, which can make you feel anxious. If you are used to drinking a lot of caffeine and then suddenly stop, you can get caffeine withdrawal symptoms, like headaches and crankiness.

Since caffeine is not a necessary nutrient in the diet and there are so many negative health effects, it is best to limit caffeine in your diet or avoid it entirely. You may think you need a little energy boost, but when the caffeine wears off, your energy level comes crashing down, making you feel sluggish, nervous, and uncomfortable. Children and teens who are used to drinking caffeine get in the habit of drinking more and more of it and end up as adults who are addicted to large amounts, making it even more difficult to stop. There has been a nationwide movement to remove soda machines from schools, making access to caffeine restricted in an effort to get children to decrease their consumption. Many adults are also trying to limit their caffeine consumption to avoid its ill effects.

Can eating chicken soup really cure a cold?

Colds are caused by viruses. When you get a cold, the virus stimulates your body's immune system to produce inflammation, which results in a cough, congestion, and nasal stuffiness. Chicken soup may help block this inflammatory reaction in several ways. First, scientists have studied chicken soup in relation to the immune system, specifically your white blood cells, which cause inflammation. Chicken soup appears to inhibit the action of certain white blood cells called neutrophils, thus decreasing inflammation. Second, some of the commonly used vegetables in chicken soup, like celery and onions, have antioxidant properties, which also block inflammation. Scientific studies have shown that aromatic seasonings such as parsley and thyme can actually open nasal passages and help remove nasal congestion. Last, by inhaling the hot vapors of chicken soup, thick nasal secretions are loosened, causing them to move out of the body more quickly. All of this data support the old adage that chicken soup is good for a cold!

Does an apple a day really keep the doctor away?

If it were that simple, everyone would be eating apple pie. Here are four basic rules that you can follow to promote good health.

First, eat well. You choose what you put into your mouth. Your diet should include lots of fruits, vegetables, and whole grain foods. A good rule of thumb is to look at the food before you eat it—if you recognize where it came from, it's probably healthy. So if you were to look at a piece of bread and could see the whole grains, you would know that the bread came from wheat. If you were to look at a piece of chocolate cake, it would be harder to see what the ingredients were, and therefore chocolate cake is not as healthy for you.

Second, get a good night's sleep. Most preteens and teens need about nine hours of sleep per night, but some children need more and some need less. As you become a teenager, your internal clock tends to shift, so that you want to stay up later at night and sleep longer in the morning. Adults need roughly seven to eight hours, although there are always individual variations. You know what you are like when you don't get enough sleep—crabby, irritable, and not as sharp as you could be in the classroom or office.

Third, exercise every day. For most kids, this means about an hour of aerobic activity every single day. For adults it's about thirty minutes a day, five days a week. You can play a sport, dance, or take the dog for a brisk walk. It doesn't matter what you do, just that you exercise enough to make your heart pump faster.

Last, make good choices—don't smoke, avoid alcohol, and don't do drugs. This advice applies to children and adults equally. If you make good choices, you will be less likely to have an accident. Don't drive if you've been drinking or get in a car with a driver who is drunk. Drugs and smoking are addictive and can have serious health consequences.

You have the power to make healthy choices that will last a lifetime. If you start while you are young, you will develop routines that will help you feel great for decades to come.

Do you really need to wait an hour after eating before going swimming?

When you eat, your body directs blood flow to the stomach and intestines to aid in the digestive process. If you eat a really large meal (like at Thanksgiving), more blood is directed to the intestines. If you then perform vigorous exercises, the body's muscles in your arms and legs need that blood flow, but they may not get it, potentially resulting in a cramp, which could be dangerous while swimming. If you eat an average meal, however, cramping is not really a problem. Many athletes eat before races or matches with no ill effect. Common sense should prevail here—if you eat so much that you feel heavy, bloated, and sleepy, swimming is not a good idea!

Bonus Body Trivia

this extra part covers some basic medical questions—from the smallest muscle in your body to the longest. After you read this part, you'll know just how many bones are in your body and which one is the tiniest.

How does an X-ray work?

X-rays are actually a type of energy wave called electromagnetic radiation. These waves pass through your body and make an image of what you look like inside. Bones and teeth appear white on X-rays. Air (like the air inside your lungs) looks black, and all of your organs (like your heart, gut, liver, etc.) look gray. Some common types of X-rays include dental X-rays to look at your teeth for cavities, X-rays of your arm or leg to check for broken bones, and X-rays of your breast (mammograms) to screen for breast cancer. X-rays don't hurt, and they can help your doctor figure out if anything is wrong with you.

What does an IQ test measure?

the term IQ stands for intelligence quotient. it is a standardized test that measures your intelligence and your cognitive abilities (meaning how you learn or acquire knowledge). The test is called the Stanford-Binet intelligence scale; it was first introduced around the early 1900s and then was modernized to the test we have today. The IQ test can be given to children or adults to compare them with other people of their same age. The test measures areas that have to do with language, memory, visual learning, and math/numbers learning. It is supposed to eliminate bias based on social or cultural distinctions. For example, the IQ test does not ask specific, learned facts, like what is the capital of Idaho? Some children may have learned that Boise is the capital, but other children, going to a different school, may not have covered this fact. No test is perfect, and the IQ test is meant to provide an overall basis for comparison. Thus the IQ test can help identify children who need extra help in school or children who would benefit from more challenging work in the classroom. Additional testing may be required to refine the results.

How much does your brain weigh?

||||||||||||||||||||||||||||||||||||||||||||||||||||||||||||||||||||||||||||||||||||||||||||||||||||||||||

an adult human brain weighs around three pounds. Most of that weight (80 percent) is water, the rest is mainly brain cells, called neurons. The bottom part of your brain, the brain stem, controls your breathing, sleeping, and heartbeat. Other animals, such as reptiles and birds, have a very similar brain stem, because all living beings need to perform the basic functions of life, like breathing. The upper part of the brain, the cerebrum (the wrinkled, wavy, squiggly part), is responsible for thought and language. Because humans have a proportionally larger cerebral part of the brain than animals, we are smarter than other animals, including other mammals.

• • • • • • • •

How many bones are in your body?

there are a total of 206 bones in your body. The smallest is in your ear and helps transmit sound vibrations so that you can hear properly. The largest, the femur, is in your thigh. There are eight bones in each wrist and five in each hand. Your ten fingers are made up of twenty-eight finger bones. You have twelve pairs of ribs and thirty-three vertebrae, which are your backbones. One of the most frequently broken bones is the arm bone, just above the wrist. One common arm fracture is a Colles' fracture, which typically occurs when you trip and try to break your fall by putting out your hand. When you land on your hand, it breaks your wrist and requires a cast to realign the bone.

What is the smallest bone in your body?

the smallest bone in your body is the stapes bone in your middle ear. The middle ear is located between the external ear (the part of your ear that you can see) and the inner ear, which is responsible for hearing and balance. There are three tiny bones in the middle ear. First, the malleus (which is Latin for "hammer") is attached to the eardrum. When sound vibrations hit the eardrum, they cause this hammer-shaped bone to strike the second bone, called the incus (Latin for "anvil"), and finally the stapes (Latin for "stirrup"—because it is shaped like the stirrup on a horse's saddle). Because the stapes is connected to the inner ear, sound vibrations are transferred from the external ear, through the middle ear, and into the inner ear, which, in turn, is connected to the fluid surrounding the brain. The stapes is only about a tenth of an inch long, but it is extremely important. Without this tiny bone, you would not be able to hear properly.

What is the smallest muscle in your body?

||||||||||||||||||||||||||||||||||||||||||||||||||||||||||||||||||||||||||||||||||||||||

the smallest muscle in the body is the stapedius, a muscle in the middle ear that is attached to the stapes bone. It is much smaller than an inch. With other tiny middle ear muscles, it protects your ear from loud noises. If you are at a concert with loud music playing, large sound vibrations beat on the eardrum. The middle-ear muscles dampen or tone down these vibrations as the middle-ear bones transmit them to the inner ear. If the stapedius muscle is paralyzed for some reason, all sounds seem too loud, as the vibrations pass through the middle ear without being toned down. This condition is called hyperacusia.

What is the longest muscle in your body?

he sartorius muscle is the longest muscle in your body. It is in your thigh and runs from the hip to the knee. The sartorius helps you to move your leg outward and rotate it sideways. You also need it in order to sit cross-legged. Sartorius comes from the Latin word *sartor,* which means "tailor"; a tailor often sits cross-legged on the floor to pin dress hems or to cuff pants.

How many times a day do you blink?

Your eyes blink spontaneously to spread tears across the surface and clean away any dirt. Two muscles are responsible for blinking: the orbicularis oculi are the circular muscles that surround the eyes and cause them to close. The levator palpebrae are the tiny muscles in the upper eyelids that cause the eyes to open.

Babies don't blink a lot, but as children get older, they blink more. You blink the most times per minute around age seven or eight and then as you become an adult you blink slightly less. The average is twelve to twenty blinks a minute or one blink every three to five seconds. Your activity level also determines how often you blink. For example, if you are reading, you blink less than when you are playing sports. So the average person blinks about 15,000 times a day!

How many times a day does your heart beat?

a t rest, a normal heart rate is between sixty and one hundred beats per minute. If your heart rate is slower than that, it is called bradycardia, and if it is faster, it is called tachycardia. When you exercise, your heart beats much faster in order to pump out more blood to your body and supply it with fresh oxygen. Your maximum heart rate is the fastest rate your heart can go. For adults, you can subtract your age from the number 220 to find your maximum heart rate. Your target heart rate during exercise should be between 60 and 85 percent of this number. If you don't like math and want to know if you're exercising at the right level, try to sing a song while you are working out. If you are exercising at the right intensity, you shouldn't be able to sing out loud, but you should be able to speak in brief phrases in between heavy breathing. The average person's heart will beat more than 110,000 times in a day. That's a lot of work!